AIRFIX
magazine guide 4

Napoleonic Wargaming

Bruce Quarrie

Patrick Stephens Ltd
in association with Airfix Products Ltd

First published - October 1974
Reprinted - 1977

ISBN 0 85059 178 3

Cover design by Ian Heath

Text set in 8 on 9pt Univers Medium by Stevenage Printing Limited, Stevenage.
Printed in Great Britain on Fineblade cartridge 90 gsm and bound by The Garden City Press, Letchworth, Herts.
Published by Patrick Stephens Limited, Bar Hill, Cambridge, CB3 8EL, in association with Airfix Products Limited, London SW18.

Contents

Editor's introduction 4
one **First things first** 5
two **Napoleonic army organisation** 11
three **Into battle!** 25
four **Napoleonic strategy and tactics** 38
five **Organising a wargames campaign** 44
appendix one **Playing rules** 49
appendix two **Miscellaneous** 63

Acknowledgements

Several people have assisted greatly in the preparation of this book, and I would like to give special thanks to Steve Tulk of the North London Wargames Group for devising the rules and for the loan of many of the figures featured in the illustrations. Ron Morris and Steve Benfield of the same club have also offered much invaluable advice, as well as serving as erstwhile opponents across the table on many occasions! I am also indebted to Richard Neibert of Der Kriegspielers in Ohio, America, for his assistance on Napoleonic organisation; to Ian Heath for doing his usual marvellous job on the drawings; Donald Featherstone and Frank Hinchliffe for the loan of photographs; and to my wife Linda (an ardent wargames pacifist!) for her patient encouragement.

Editor's introduction

Wargaming–the art of recreating in miniature the great battles of history through the use of model soldiers–is probably the fastest-growing pastime in Britain today. And amongst wargamers, the Napoleonic era is by far the most popular.

The reasons for this are legion : it was a colourful period, when the armies almost seemed to vie in a perpetual fashion parade of the most extravagant uniforms imaginable; it was a significant period historically in terms of tactical and strategic innovation; at one time or another it involved practically every country in Europe, and many outside as well; and it saw the rise of some of the most outstanding military leaders of all time. Moreover, being so popular, it is also one of the best documented historical periods so far as information on the arms, uniforms, battles and tactics is concerned.

For these reasons it is also a very confusing field of study for the newcomer, and one of the aims of this book has been to collate the basic information any beginner needs to set him on the right footing for a more detailed study. At the same time it includes all the essential information he needs to begin organising and fighting with a miniature Napoleonic army.

The contents include discussion of the basic wargames rules of time, space and scale; organising a Napoleonic army; national characteristics of the combatant powers; leadership characteristics of the most important generals; Napoleonic strategy and tactics in real life and on the tabletop; and a complete set of playing rules based on those used successfully over a long period of time by the North London Napoleonic Wargames Sub-Group. Finally, there are lists of model soldier manufacturers and their addresses, and a bibliography of useful books for further reference.

My basic aim has been to answer all the main questions which puzzled me when I began Napoleonic wargaming some years ago, and I shall be well satisfied if this book does the same for you. Readers who wish to take their Napoleonic wargaming further may like to read my other book on the subject, *Napoleon's Campaigns in Miniature* (also published by Patrick Stephens Ltd), which examines these questions and many others – such as recruitment, training, supplies and sieges – in greater detail.

BRUCE QUARRIE

one

First things first

If you are reading this book at all the chances are that you are either already a Napoleonic wargamer, or else interested in becoming one. This chapter is really addressed to all those readers who would like to begin wargaming but haven't yet put their toes in the water: those of you who do know something about the subject may prefer to move straight on to chapter two, unless you are uncertain of the basic rules of scale, time and space.

There are few things quite as exciting—to the enthusiast—as a Napoleonic battle. Brightly uniformed infantrymen march and countermarch across the green tabletop; little puffs of cotton wool 'smoke' appear opposite the muzzles of the guns and figures topple to the ground; ponderously, a regiment of heavy cavalry picks up speed and charges towards a battalion of infantry which is frantically trying to form a square; a unit breaks and flees in disorder, gaily plumaged hussars in hot pursuit. And so it goes on: all the colour and a lot of the excitement of a real battle, but with one important element missing—blood. And at the end of the evening, instead of one combatant being exiled to lonely St Helena, the players pack up their miniature armies and return to their homes, all the best of friends...until the next encounter!

How does it all work, and how do you begin? Well, this book I hope will explain in more detail how it works, but without getting involved at this stage, wargaming involves the use of small plastic or metal model soldiers and guns, all correctly (we hope!) painted in the uniforms of the period. These are manoeuvred across the tabletop with the help of tape measures on which a centimetre equals ten yards of real ground. With the aid of the same tape measure the guns and muskets fire at targets which are within their range, and casualties are calculated according to carefully worked out tables. A unit is threatened—its morale is assessed according to a further set of tables. A melée, or hand-to-hand fight is conducted in the same way, everything relies on a set of rules.

However, the first thing that you have to understand is that practically nobody places any reliance on any other player's set of rules. There are *no* standard moves or responses as in Chess. Everything relies on historical and personal interpretation of what actually took place during the first turbulent 15 years of the 19th Century.

But if nobody agrees on the rules, I can hear you asking, how on earth does anyone ever manage to fight a battle at all? The answer is compromise. If two enthusiasts each have their own set of rules, then one set is used at one player's home, the other at the second player's. Similarly, in a wargames club, everybody will initially have their own pet rules. Gradually, through playing experience, these will become modified, and one set will emerge as more popular than others. Eventually this will be adopted as the club set of rules. And when it comes to inter-club competitions, much the same situation applies as with individuals: the guests fight to the host club's rules.

Despite all I have said though, a degree of standardisation is gradually creeping in. Through long debate in the wargames magazines, such as *Wargamer's Newsletter* and *Battle*; through inter-club correspondence; through research and discussion by and between individuals, the main elements of Napoleonic warfare are being codified and gradually agreed.

The rules discussed and presented in this book are not definitive—no rules ever are. However, they are based on several years' playing, first as a solo wargamer, later with a small group of friends, and then as a member of one of the country's largest Napoleonic groups, the Napoleonic Sub-Group of

Napoleonic wargame in progress with the Prussians running up against French guns in the centre and a major attack developing in the foreground (Frank Hinchliffe).

the North London Wargames Group. In many essentials, these rules are based on those used by the Group, but I am not completely partisan and they also owe much to the rules devised and used by other clubs up and down the country, as well as to my own research and thoughts.

Some readers will shoot them down as being too complicated; others will say I've missed a lot out. When you are dealing with a topic like this you cannot possibly please everyone, and to those who are tempted to argue, I would just like to say–give these rules a chance. Try them over half a dozen battles. If you still don't like them you can always go back to your old ones.

All right–now we've got a set of rules. Perhaps not everybody likes them, but they will do for the moment. What else do you need to begin playing?

The first requirement, obviously, is an army, which need not be as expensive as you might fear. Assuming that you intend to use the popular 20/25 mm figures rather than one of the 'odd' scales such as 5, 15 or 30 mm, the choice is vast. Airfix now manufacture quite a large range of Napoleonic figures, all Waterloo period French and British at the moment unfortunately, but the range is being extended all the time. These sets allow you to build up quite large units at very little cost, and details of how to organise your regiments are given in the next chapter.

But do, please do, make some attempt to paint them. Even if your hand isn't as steady as you would like, it isn't too much to ask of anybody a black shako, red or blue jacket, and grey or white trousers, with perhaps a touch of pink for face and hands, and black shoes and musket–is it? If painted in batches of a dozen or so at a time, doing all the hats first, then all the jackets, and so on, it does not take long, and the result in terms of tabletop appearance well justifies the slight effort.

If you have a little more money, or particularly fancy something other than a French or British army, there are several specialist firms manufacturing wide ranges of Austrian, Prussian, Russian, Spanish, Polish, Bavarian and a host of other figure ranges for this period. A list of these manufacturers is given later. Naturally, metal figures are more expensive – up to 15p for a foot figure–but their extra weight makes them attractive to handle, paint adheres to them better than it does to the polythene plastic made by Airfix, and there is a much wider choice. Even if you begin by using the cheaper figures, you will probably want some personality figures to represent your favourite generals, and for this you will have to invest in at least a few metal figures.

Right! Now you've got a set of rules—this book—and some painted figures. What else is required? First is a tape measure or extendable steel rule (available from hardware and D-I-Y shops) marked out in millimetres and centimetres, since as you will see later we are going to use a basic scale of 1 mm to 1 yard.

The next requirement is a dice marked 1,1,2,2,3,3 rather than the usual 1,2,3,4, 5,6; and two other dice marked 2,3,3,4, 4,5. If you cannot find any of these

(called 'fire dice' and 'average dice' respectively) in your local shops, they can easily be made by painting out the dots on three ordinary dice and painting in the new numbers. The reasons for using these specially marked dice will become clear later.

Having assembled all these bits and pieces, your final requirement is a playing area and some terrain. The playing area can be an ordinary dining room table commandeered temporarily for a battle, or an elaborate affair 18 feet long and six feet wide covered with green baize or felt. Generally speaking, most action in a battle takes place in a fairly limited area of the table, so size is not *all* that important, but if you are contemplating a really large table, remember that it should be no wider than twice the reach of your arm, or you will not be able to move anything in the centre!

If a proper table is beyond you and your dining room table is really too small, a simple solution is to buy a couple of pieces of ¾" chipboard, say four feet square, paint one side of each green, and lay them over the top of your dining room table, making it much bigger. This, in fact, is what I have done successfully for years.

If your parents or wife object to possible scratches on the dining table from the chipboard, lay some sheets of newspaper down first to protect it. Domestic harmony is very important, especially since most (female) friends and relatives seem unable to understand our preoccupation with 'playing with toy soldiers'!

So now you have got a bare playing area of some sort, but battlefields are not all flat billiard tables—how about some hills? These can very simply be constructed by buying or begging a few polystyrene ceiling tiles, cutting them into irregular shapes, and placing them on top of one another to form low 'stepped' hills. The pieces should, of course, be painted the same colour as your baseboard, and in this context it is worth noting that enamel or other oil-bound paints tend to dissolve polystyrene, so a poster or emulsion paint should always be used.

Next, rivers and roads can be applied either by using coloured chalk, or by cutting out lengths of blue and brown material from any scraps you can beg around the home or from the local fabric shop. Alternatively, some very nice vacuum-moulded river lengths are manufactured by Micro-Mold, and can be painted blue to represent rivers or brown to represent roads.

Cornfields and areas of differently textured ground can be shown by using some of the small carpet sample squares often seen given away or sold cheaply by carpet shops.

Trees and hedges can be bought from model railway shops, but tend to be expensive so it is worthwhile making your own. Trees can be made from twigs stuck in Plasticine bases, with cotton wool dyed green to represent foliage, while hedges can be made from cotton wool or from the remains of the ceiling tiles you have already used in constructing your hills. Walls can be made in the same way, or from Plasticine or moulded from Plaster of Paris.

Houses and other buildings are available in many forms commercially, varying from the few still left in the Airfix Trackside range to the beautifully detailed—and expensive—German imports designed to accompany model railway layouts. There are also a wide variety of cardboard cut-out buildings available from most good model railway stockists, or you could try making your own from cardboard and/or plastic card.

I think we have got everything we need now, so let us consider the three things which make wargaming possible: the rules of

Space, time and scale

One fact which becomes apparent the moment you start thinking about organising a wargames army is that it would take an awfully long time to build up a reasonable force on a one-to-one basis. The average infantry company would consist of a hundred or so model soldiers, a battalion four hundred to a thousand, and so on. Apart from the expense involved in building up such a force, and the time needed to paint it, it would also take up so much room that there would be no possibility of tabletop manoeuvre at all.

For this reason we must adopt a ratio which scales down the sizes of our units to reasonable proportions. The most commonly employed ratio is 33:1 (ie, one model soldier *represents* 33 actual soldiers) and, since I am assuming you want to get your wargaming off on the right foot, this is the ratio we shall use. It gives us an average infantry company of three model figures, a battalion of 24 or so, and a cavalry squadron of three to five, but precise details are given in the next chapter.

Similarly, since an artillery battery usually consisted of six or more guns, we scale this down and allow one model gun to represent a complete battery.

The next problem is that of ground scale. If this was based on the actual size of our model soldiers (25 mm) we would have a scale of roughly 12½ mm to one yard, 125 mm to ten yards and so on. This would mean that the average wargames table, five feet wide by eight feet long (1,524 mm x 2,438 mm) would represent a ground area only 122 by 195 yards, while our wargames company of three figures (99 men) would occupy a frontage of around five yards! Thus, in the same way that we had to scale down the figures, so we must scale down the terrain, and the usual ratio used here is one millimetre to one yard. This makes our average wargames table a more respectable 1,500-odd yards wide and 2,400-odd yards long.

The next question is that of time, or, in other words, how much actual battle time does a Move on the tabletop represent? Once again, there is a standard formula which we shall use, this being that one Move represents two and a half minutes of actual time. This is a handy scale since it allows the troops to move into contact fairly quickly, speeding up the game, yet does not permit them, for example, to proceed from one end of the table to another in a single Move as could happen if you had a scale of, say, half an hour to a Move.

In 2½ minutes the average infantryman could move 125 yards (slightly less than one yard per second) or reload and fire his musket five times. Not all troops were as mobile as others though, and this is reflected in the rules by giving different types different movement rates. It is obvious, for example, that light infantry acting as skirmishers could move more quickly than line infantry who were having to maintain a close formation, or that cavalry could move more quickly than infantry, and so on.

Several factors emerge from our rules for time and space, one of the most important being terrain. Obviously, if one model soldier is scaled down to represent 33 actual men, it would be silly to have one model house or tree representing only one actual house or tree, so these must be scaled down in the same ratio. Thus three or four model trees becomes a respectable wood, while half a dozen model houses is a sizeable town.

And what about vertical scale? Having adopted our ground scale of 1 mm: 1 yard, our wargames soldiers are now giants 25 yards in height, while houses become skyscrapers several hundred feet tall. Once again we compromise and lay down some standard rules. These are important when we come to consider factors such as the amount of dead ground behind a house, wood or hill. Thus all houses are said to be 40 feet high; all trees are the same; and one contour of a hill is also 40 feet.

Having covered the basics, there remains just one other very important

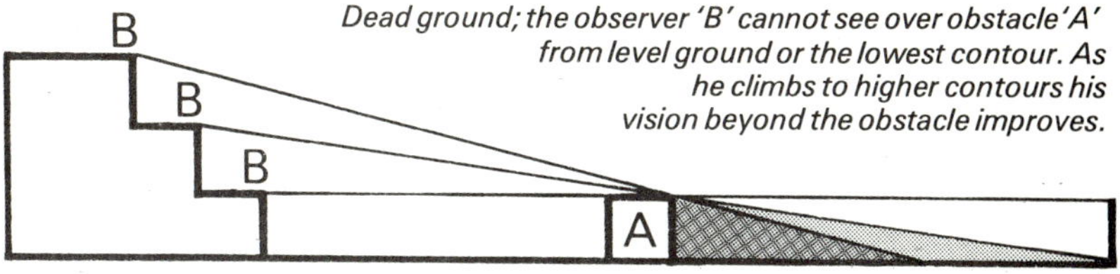

Dead ground; the observer 'B' cannot see over obstacle 'A' from level ground or the lowest contour. As he climbs to higher contours his vision beyond the obstacle improves.

factor related to scale to consider, and that is the frontage of our wargames companies and battalions.

Most wargamers mount their figures on rectangular card bases for ease of handling and moving, and it is very important that these are all the same size. The reason for this is obvious if you consider the situation where wargamer A has crammed the three figures in each of his companies onto a base 25 mm wide, while wargamer B has allowed his men more elbow room and fixed them to bases 35 mm wide. Obviously, in a melée situation, A will be able to squeeze four of his own figures up against every three of B's. This is not only unfair, it is also unrealistic, since the intervals between men were fairly standardised in all armies of this period.

Taking the French as the average, therefore, with a company frontage of roughly 30 yards when drawn up in line three ranks deep, we arrive at a standard three-figure base 30 mm wide. A two-figure company would therefore have a base 20 mm wide; a four-figure company one 40mm wide and a five-figure company one 50 mm wide.

Note that when fixing your figures to their bases one figure in three should be left detachable so that casualties can be removed. By the same token, when troops are acting in open order as skirmishers, the one individual figure in three can be removed from its base and moved separately, representing the extended frontage of this formation.

Cavalry should also be mounted on squadron bases. Heavy cavalry rode virtually boot-to-boot to emphasise their shock effect, and two yards per man would be an average frontage. Thus a two-figure squadron should have a base frontage of 40 mm, a three-figure squadron of 60 mm, four figures 80 mm and so on. Light cavalry tended to ride in a much looser formation and this is represented by making the base frontage of each light cavalry figure 25 instead of 20 mm, thus

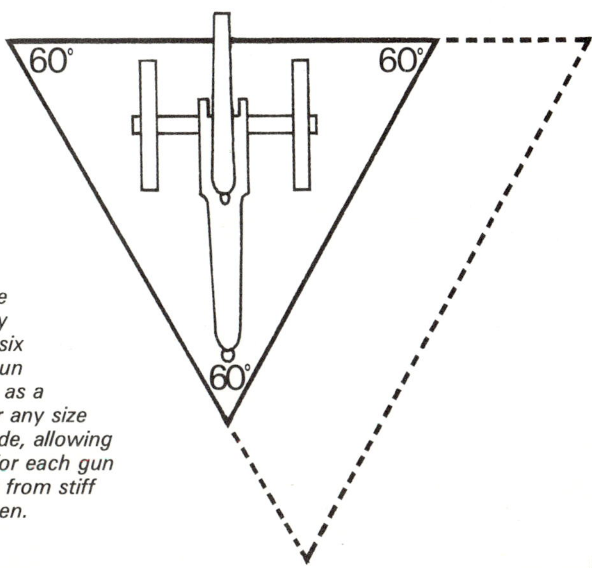

Same size template showing an artillery battery base for a six or eight (dotted) gun battery. Using this as a basis templates for any size battery can be made, allowing 1 cm of frontage for each gun in the battery. Cut from stiff card and paint green.

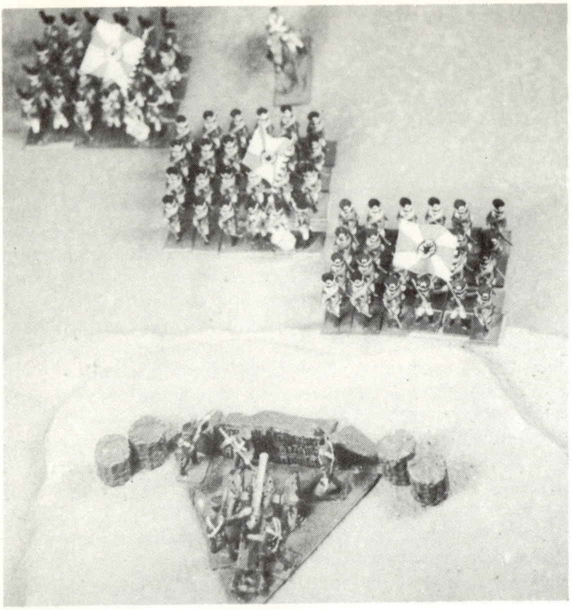

A French 12pdr battery on its triangular base, emplaced on a polystyrene ceiling tile 'hill' behind earthworks and gabions, is assaulted by Russian columns moving up in echelon formation.

50 mm for a two-figure squadron, 100 mm for a four-figure squadron, etc.

Finally, since our model guns represent actual batteries of artillery, we must also mount these on bases when in action so as to depict the battery frontage. The average interval between guns was ten yards, so a six-gun battery should have a base 60 mm wide, eight guns 80 mm, and so on up to the massive Russian 14-gun batteries.

While the infantry and cavalry bases are simple rectangles of card, it is worthwhile making the artillery bases as equilateral triangles, with each side the same length. The front face of this then represents the battery frontage, while the angle (60°) subtended by the other two sides shows the battery's field of fire without moving the actual battery front: in other words, the guns can shoot at any target within the larger triangle formed by extending the sides of their base. The diagram on page 34 makes this clear.

Having established our basic rules of time and scale, let us move on to consider the basic organisation of the main Napoleonic armies, and how they can best be reproduced in our scaled-down fashion.

Polish Vistula Legion lancers advance down a Spanish street during a Peninsula War game (Donald Featherstone).

two

Napoleonic army organisation

The basic organisation of all armies during this period was very similar, although there were differences which can have quite a marked effect on tabletop tactics. The three main arms were the infantry, cavalry and artillery, together with supporting services.

Infantry came in various types: the bulk of the men were line infantry, variously known as fusiliers or musketeers. They were organised in companies of anything between 50 and 200 men. Companies were grouped together in battalions, usually between four and ten companies making up the battalion. In most armies two of these companies were composed of élite troops, one being grenadiers (selected from the tallest men in the battalion), the other light infantry (variously called tirailleurs, voltigeurs, etc), who were specially trained to operate in a skirmish line.

Apart from the light infantry companies in the Line battalions, there were also in many armies separate battalions of light troops. These were organised on the same basis as the Line battalions, but the men of the 'fusilier' or centre companies were usually called chasseurs (or jägers) while the élite companies were carabiniers (instead of grenadiers) and voltigeurs.

Some armies also concentrated all their grenadier companies to form grenadier battalions, regiments and Divisions.

Apart from the regular Line battalions, there were also in most armies units of Guard troops, who were generally organised in much the same way but did not, of course, have any élite companies since the whole battalion was élite.

Battalions in their turn were organised into regiments, with between one and six battalions per regiment. One of these battalions was usually a depôt unit, however, used to supply replacements to the field battalions, and for garrison duties.

Regiments were grouped into brigades of two to six regiments; brigades into Divisions, usually together with some cavalry and artillery; and Divisions into Corps, which were essentially miniature armies in their own right, containing a balanced mixture of all arms together with their supporting services—sappers, pontonniers, miners, gendarmes and HQ staff.

Cavalry consisted of light and heavy troops, the former including hussars, chasseurs à cheval, light dragoons and lancers, the latter including cuirassiers, carabiniers, dragoons and grenadiers à cheval. The light cavalry was mainly used for scouting and escort duties and pursuit after a battle, the heavy cavalry for shock effect on the battlefield. Lancers, however, were also used for shock effect since their weapon made up for their lack of weight.

The basic unit of cavalry organisation was the squadron, which corresponded to the infantry company, and also consisted of 50 to 200 men. The squadrons were in turn grouped into regiments of between four and ten squadrons, while several regiments were frequently brigaded together into larger brigades or Divisions for shock effect on the battlefield.

There were two types of artillery in most armies, foot and horse. The foot artillery batteries contained the heavier guns and moved at the same pace as the infantry, while the horse artillery usually had lighter guns and whose gunners, being either mounted or riding on the limbers, moved at cavalry speed.

The basic unit of artillery organisation was the battery, consisting of six to 14 guns, frequently including two or more howitzers. A gun was crewed by between ten and 30 gunners. In some armies individual guns or small groups of guns were allocated to each infantry regiment, while in others the batteries were themselves grouped into regiments which could act to put down a heavy fire against a selected point on the enemy line.

However, there were wide differences between the main armies of the period and it is worth now considering the individual characteristics and organisation of these in turn.

In some cases the actual sizes of battalions or squadrons has been impossible to scale down appropriately without giving an unwieldy number of figures per unit, or alternatively, without breaking away from the correct company organisation. In these cases the company or squadron strength has been rounded up or down to the nearest figure (33 men) so that the organisation is at least correct even if the total number of men per battalion or regiment is slightly over or under strength.

1 The French Army

Born in the fervour of the French Revolution and tempered in the fires of the Revolutionary Wars, the French Army was a highly strung, flexible and mobile force by the time Napoleon assumed full powers over France. Full of boundless confidence in its ideals, it was cheerful and aggressive in the attack, despondent but stubborn in defence, and capable of tactical and strategic miracles when well led.

In the years before 1807 the French Army was organised as follows: the line infantry into regiments of three field and one depôt battalion; the cavalry into regiments of four squadrons; and the artillery into regiments of varying numbers of batteries.

Each battalion consisted of nine companies, seven fusilier, one grenadier and one voltigeur (light). The average battalion strength was 7-1100 men which, using our wargames men: figure ratio of 33:1, gives an average battalion of 27 figures, organised as three grenadiers, three voltigeurs and 21 fusiliers. The light infantry battalions were organised exactly the same, except that the 'grenadiers' were called carabiniers, and the 'fusiliers' chasseurs; voltigeurs remained voltigeurs.

The average strength of a cavalry squadron was between 140 and 180 men, which gives us a wargames unit of five figures to a squadron, 20 to a regiment.

An artillery regiment usually contained 22 batteries, but since it is very rarely that one sees this number of guns employed on the tabletop, we must concentrate our attention on the battery organisation.

The French used a variety of guns, ranging from tiny 4pdrs to massive 12pdrs (excluding siege artillery), and a number of howitzers, mainly 5.5" (10lb). Foot artillery batteries consisted of eight pieces, 6 x 8pdrs and 2 x 5.5" howitzers: the 12pdrs were organised

French infantry brigade, with voltigeur companies in skirmish order to the fore and columns behind (Donald Featherstone and R. Tennant).

separately into Reserves and later amalgamated with the Guard. Line horse artillery batteries consisted of 6 x 4pdrs, Guard horse batteries of 6 x 6pdrs (usually captured). Each gun was crewed by 12-18 men, giving approximately 70-120 men to a battery excluding drivers and other 'extras'.

On the wargames table a battery of six or eight guns is generally represented by a single model gun, crewed by 2-4 figures, although some wargamers use as many as five or six figures per gun.

The Imperial Guard infantry were organised in two battalions per regiment, each battalion consisting of eight companies of 80 men, giving 16 wargame figures per battalion with two figures per company. The Guard cavalry was organised on exactly the same basis as the Line, except that the Chasseurs à Cheval always had five squadrons plus the attached squadron of Mamelukes.

In 1805 the French field Army consisted of the following: 89 line infantry and 26 light infantry regiments; 30 regiments of dragoons, 12 of cuirassiers, two of carabiniers, ten of hussars and 24 of chasseurs à cheval; eight foot artillery and six horse artillery regiments; plus sappers, siege batteries, pontonniers, miners and gendarmes. The Guard at this time comprised simply the two regiments of Foot Grenadiers and Chasseurs, a regiment of Marines, plus the Grenadiers and Chasseurs à Cheval, Mamelukes and Gendarmes d'Elite.

In 1806 a regiment of dragoons–later named the Empress Dragoons–was added to the Guard, together with a horse artillery regiment.

During the years 1807-9 several major changes and additions were made. First of all, the line and light infantry battalions were reorganised into six instead of the previous nine companies, while the average battalion strength also dropped. From 1807 until the end of the period therefore, wargames infantry battalions are best organised as 24 figures, comprising four grenadiers, 16 fusiliers and four voltigeurs, or four carabiniers, 16 chasseurs and four voltigeurs in a light battalion.

After the 1812 debacle these numbers were seriously reduced, and a battalion for the campaigns of 1813-14 should consist of merely 21 figures, three to a company. For the 1815 Waterloo campaign, however, a return to 24 figures is appropriate.

Several additions were made to the Guard during the reorganisation of 1807-9. Second regiments of Grenadiers and Chasseurs were formed, organised the same as the parent regiments. These years also saw the beginning of the Young Guard, when the 1st and 2nd regiments of Tirailleur Grenadiers, of Tirailleur Chasseurs and of Conscript Grenadiers and Chasseurs were formed. Each of these consisted of four companies of 120 men, giving a wargames strength of 16 figures per battalion. A regiment of Polish lancers and the lancers of Berg were added to the Guard cavalry strength, while a Guard foot artillery regiment consisting initially of four batteries of 12pdrs and 8'' howitzers was formed in 1808.

Further changes occurred during 1810 and 1811, increasing the French Army to the following by 1812, at the start of the Russian Campaign: 105 line infantry and 33 light infantry regiments; 14 cuirassier, two carabinier, 24 dragoon, 13 hussar, 28 chasseur à cheval and seven lancer regiments, plus artillery as in 1805. The Guard saw the addition of a third (Dutch) regiment to the Grenadiers, regiments of Dutch and Lithuanian lancers, and the strength of the foot artillery increased from four to eight (1810) then nine batteries. The Young Guard regiments changed their names and were increased to 12 strong, as: 1st-6th Regiments Tirailleurs and 1st-6th Regiments Voltigeurs, each regiment of two battalions of four companies as before.

French casualties during the invasion of Russia were very high, and Napoleon really had his work cut out to reform an effective striking force in order to fight back against the Allies in 1813. At the beginning of the campaign he could muster 132 line and light infantry regiments, most of them seriously under strength as mentioned above, and 22 of them formed from the old National Guard cohorts. The cavalry was in a particularly bad way, most regiments being able to field no more than three squadrons, and most of those under strength, so for wargames purposes a squadron of four figures, with 12 figures per regiment, is about right.

The Guard remained virtually unchanged except that the Dutch Grenadiers and Lithuanian lancers had been lost in Russia, while the lancers of Berg defected to the Allies after the battle of Leipzig in 1813. Three regiments of Guard scout lancers were formed, however, together with a short-lived Young Guard artillery regiment which was reduced to the Line in 1814.

When Napoleon went into exile in 1814 the only troops he was allowed to take with him were six companies of Grenadiers and Chasseurs of the Guard, a squadron of Polish lancers, and of course his personal bodyguard, the five squadrons of Chasseurs à Cheval. On his return from Elba however he wasted no time in calling everybody he could think of back to the colours, with the result that, by the beginning of the Waterloo campaign, he was able to field the following: 90 line, 15 light and four Swiss regiments of infantry; 20 regiments of dragoons, ten of hussars, 12 cuirassiers, two carabiniers, six lancers and 12 chasseurs à cheval, plus approximately 40 foot and 40 horse artillery batteries. The Imperial Guard consisted of the four Old Guard Grenadier and Chasseur Regiments, the two Middle Guard regiments, four regiments of voltigeurs and four of tirailleurs in the Young Guard, the Polish lancers, Dutch lancers, Empress Dragoons, Grenadiers à Cheval and two regiments of Chasseurs a Cheval. Not all saw action at Waterloo, however.

It must be remembered that through most of this period the French Army was considerably reinforced by allied contingents, especially between 1807 and 1813: during the invasion of Russia, for example, no less than six out of 11 Corps were Allied, while the remaining five included only three which were essentially French! These allied armies are a very fruitful field of study for the wargamer, since the uniforms were extremely colourful while the size of the armies was generally small, meaning that a wargamer could recreate in miniature, for example, the entire Saxon or Westphalian army, to name but two. However, in a book of this size it is not possible to go into all of these, and anyone seeking further information is advised to consult the Casterman book *Arms and Uniforms of the First Empire* Volume 2, by Liliane and Fred Funcken, or *Napoleonic Organisation,* The Wargamers' Library volumes II and III, which are available from Der Kriegspielers, 6335 Rosa Linda Drive, Dayton, Ohio 45459, USA, price $4 each including airmail postage.

2 The Austrian Army

Since the French fought more battles against the Austrians than any other nation during the Napoleonic Wars, it is only fair to consider them next.

Pre-1805

Up to 1805 the Austro-Hungarian Army consisted of the following: 63 regiments of line infantry, 17 of Croatian light infantry and one of jägers; 21 battalions of grenadiers; 12 hussar, six dragoon, six chevauleger, eight cuirassier and three uhlan (lancer) regiments;

The Westphalian Guard Grenadiers are just one of the many colourful units from minor states which can be incorporated in a wargames army.

An Austrian grenadier battalion on the left (24 figures) with a fusilier battalion in the centre (30 figures) and half a dozen jagers on the right.

and four regiments of artillery and supporting services.

The Austrian line regiments comprised two field and one depôt battalion, each battalion being made up of six companies of fusiliers, each company being approximately 160-strong. This gives a wargames unit of 30 figures, five to a company. The Croatian light infantry regiments were organised on the same lines, except that average company strength was only 120 men, giving a wargames unit of 24 figures, four to a company. The jäger regiment and grenadier battalions should be the same, ie, 24 figures.

Austrian cavalry was of very high quality, certainly superior to the French before 1805. Their cavalry regiments were also considerably stronger than the French. A heavy cavalry regiment normally consisted of eight squadrons of 120 men, giving a wargames strength of four figures to a squadron, 32 to a regiment, while the light cavalry usually had ten squadrons to a regiment giving 40 figures!

Austrian artillery was very poor, consisting mainly of 3pdrs which were assigned two to each infantry regiment, a few 3 and 6pdr 'horse artillery' batteries which were only slightly more mobile than foot artillery, and reserve batteries of 6 and 12pdrs. For wargames purposes assign one model 3pdr with one crew figure to every two battalions, firing as a two-gun 'battery' (see rules on page 58), one 6pdr 'horse' battery to every 20 squadrons of cavalry, firing as a six-gun battery, and one 12pdr battery of eight guns to every eight battalions.

Post-1805

After the defeat at Austerlitz a considerable reorganisation programme was undertaken in the Austrian Army. Regiments were reorganised into three field battalions, the first two of six companies as before, the third of only four companies. This should be reflected on the wargames table by making one in every three battalions only 20 figures strong.

A further seven jäger regiments were raised, organised the same as the original one; this was increased to nine in 1809.

The cavalry was reorganised as follows: heavy cavalry regiments into six squadrons instead of the previous eight, giving 24 figures to a regiment; hussars remained at ten squadrons, while all other light cavalry regiments were reduced to eight squadrons, or 32 figures on the table.

The most significant change was in the artillery; the system of assigning guns to regiments (except for the Croatian light regiments which retained theirs) was abandoned. Batteries of six 3pdrs or six to eight 8pdrs were instead assigned to every two regiments, while Position Batteries of four 12pdrs and two 7'' howitzers were allocated to each infantry Division: a Division being between two and four brigades of four or five regiments—so unless you have a minimum of 24 infantry battalions on the table you should not really have a Position Battery.

Beginning in 1808 the Austrian Army began recruiting a large number of Freikorps or Landwehr (militia) units; in general they were not intended for use in battle, but to release regular troops from garrison duties. Six battalions of Freikorps, each of six companies (30 figures), plus a company of riflemen raised in Vienna, did see action at the battles of Ebelsberg, Wagram and Aspern, where they acquitted themselves well. A further 166 Freikorps battalions were eventually raised but these saw little action and need not concern us.

In 1812 Austria, somewhat reluctantly, was forced to contribute a Corps to the French Grande Armée for the invasion of Russia; although slow on the offensive this force did sterling service on rearguard during the disastrous retreat from Moscow, and gives wargamers the chance to match Austrians against Russians for a change.

During the 1813-14 campaign the basic organisation remained as before save that the infantry regiments in the field only had two battalions, while all the cavalry regiments were two squadrons short. Interestingly, for this campaign three Light Divisions were formed, each consisting of four jäger battalions, an average of nine squadrons of light cavalry and a 6pdr horse battery; a clash between one of these and a French rearguard force would make a lively skirmish on the tabletop.

The Austrians adopted a faster march rate in the later stages of the wars, and also French-style battalion attack columns, giving them much greater manoeuvrability and tactical flexibility.

3 The Russian Army

Next to Austria, the most important Continental power arrayed against France was that of Imperial Russia, and like Hitler more than a century later, Napoleon found the 'bear' a tough nut to crack.

In 1805 the Russian Army consisted of the following: 13 grenadier, 83 musketeeer and 22 jäger regiments; 30 dragoon, six cuirassier, eight hussar, three uhlan, and one Tartar regiment;11 artillery regiments and supporting services; numerous sotnias (squadrons) of Cossacks, and the Imperial Guard. The latter comprised three infantry regiments, one battalion of jägers, two cuirassier, one hussar and one cossack regiment, plus a horse artillery battery.

The musketeer regiments each had one battalion of grenadiers and two of musketeers, while the grenadier regiments had one battalion of grenadiers and two of fusiliers (musketeers trained to operate in skirmish order as Light when required). The jäger regiments were also of three battalions. The Guard infantry regiments were organised exactly the same, with three battalions, except for the 1st (Preobrazhenski) which had four battalions.

Paper strength of the line battalions was 738, of the Guard 764, giving 22 and 23 figures respectively on the wargames table. However, since it is impossible to organise these numbers into the required four companies per battalion, most wargamers compromise by making the line battalions 20 figures strong, with five figures to a company: and the Guard battalions 24 figures strong, six to a company. Others prefer to make it 24 figures all round since the former method makes the Guard units *too* strong, and this is probably the best procedure to adopt.

The light cavalry regiments each had ten squadrons, heavy cavalry five squadrons, and the average squadron strength in the field was around 140, so a wargames squadron should be four figures.

Russian artillery was among the best in Europe, and certainly they had more of it than anybody else except the French. The artillery was organised into three different types of battery as follows: Position Batteries (heavy)– eight 12pdrs, four 18pdr licornes (a type of howitzer with a flatter trajectory), and two 9pdr licornes; 'Light' Batteries– eight 6pdrs and six 9pdr licornes; and Horse Batteries -12 6pdrs. For the sake of simplicity on the wargames table the 18pdr licornes are treated as 8" howitzers and the 9pdrs as 6" howitzers. Because of their size Russian batteries should be represented by at least two model guns, heavy batteries crewed by eight figures, 'light' by six figures and horse by four figures.Fire from each battery is then worked out in two stages (or three in the case of the heavy batteries); time consuming but horribly effective!

The 1805 Campaign

During this period the Russian Army was not organised into Divisions and Corps as other European armies, but into 'Inspections', each containing at least one grenadier plus several musketeer and jäger regiments.

After 1805, however, the Inspection system was dropped and the army was reorganised into Divisions, average composition of which was as follows: one grenadier and one jäger regiment; four musketeer regiments; two heavy and one light cavalry regiment; two heavy, three light and one horse battery. The Imperial Guard formed a Division on its own, consisting of 33 infantry battalions, 35 cavalry squadrons and 84 guns. This included the newly raised Guard uhlan and Guard dragoon cavalry regiments, as well as several musketeer and grenadier regiments.

1807

In preparation for the new campaign several new units were added, including four more line divisions (total 22 by the end of the year), six more jäger regiments, and another hussar and uhlan regiment. The Imperial Militia, first raised in 1806, acquired a second battalion and was renamed the Finnlandski Guard Jägers as part of the Guard Division.

1809-10

A further two Divisions were raised during 1809, while in 1810 a considerable reorganisation took place thoughoutthearmy. The most significant change was the reconstruction of the infantry battalions along essentially French lines. The grenadier battalions now consisted of one company of grenadiers, two of fusiliers and one of tirailleurs (fusiliers trained to act as skirmishers). Similarly, the musketeer battalions were now organised as one company of grenadiers, two of musketeers and one of tirailleurs. The jäger battalions had one company of jäger grenadiers and three of jägers.

For this period the 24-figure organisation is really essential, giving six figures to a company.

At the end of this reorganisation the Russian Army consisted of no fewer than 437 infantry battalions, 399 cavalry squadrons and 25 artillery brigades (of four to six batteries), in 26 Divisions.

In 1811 another line regiment was increased to six battalions and transferred to the Guard, while a Guard Light battery was also formed. One musketeer regiment was converted to grenadier, while a further 14 musketeer regiments were made into jägers. Four additional jäger regiments were raised (bringing the total to 50), plus 15 musketeer regiments and four marine regiments. A Russo-German Legion was also raised from German POWs, consisting of seven infantry battalions, two hussar regiments, two horse artillery regiments and one jäger battalion.

1812

Early in 1812 the system of mixed Divisions was abandoned in favour of a Corps system based closely on the French pattern. Each Corps consisted of two Divisions of three brigades. The grenadiers were massed into two Divisions, while the line infantry Divisions each consisted of four musketeer and two jäger regiments. The cavalry was also reorganised into two cuirassier Divisions of five regiments and eight cavalry Divisions, usually of four dragoon and two hussar or uhlan regiments. Two further Guard battalions were also formed, the marines and sappers, both of which fought as light infantry.

Further minor changes were incorporated in the Russian Army after 1812 but these are insignificant and the basic organisation remained the same. A further two line regiments, including the famous Pavlovski with their old-fashioned mitre hats, were incorporated into the Guard in 1813.

4 The Prussian Army

As the other major Continental power allied against France during the Napoleonic Wars, let us examine the state of Prussia before moving on to Britain. At the beginning of the period the Prussian Army was one of the worst of the 'Big Five'. Harking back to the glories of Frederick the Great, its organisation, leadership and equipment was outdated, ponderous and inflexible. By 1815, after the shock of Jena and the experience of the 1813-14 campaign, it was certainly superior to the Austrian army and in many ways the equal of the Russian.

Post 1806 Prussian infantry regiment drawn up in two-deep line, the two musketeer battalions here flanking the fusilier battalion.

In 1806 the Prussian Army consisted of: two regiments of Foot Guards, 57 regiments of musketeers, 29 battalions of grenadiers, eight brigades of fusiliers (light infantry) and three battalions of Field Jägers; 13 regiments of cuirassiers, 14 of dragoons, nine of hussars, a regiment plus a battalion of Towarozys (uhlans), one squadron of horse jägers, four regiments of foot artillery and one of horse artillery.

Organisation of the infantry was complex and is difficult to recreate accurately on the wargames table. Each regiment contained three battalions of musketeers, two companies of grenadiers, and one company of reservists. The musketeer battalions were all supposed to have four companies, but during the 1806 campaign many of them still had five, each of approximately 165 men. However, since this would give five figures to a wargames company and 25 to a battalion, most wargamers use the regulation four-company organisation giving 20 figures per battalion. A regiment should therefore consist of 60 musketeer figures, ten grenadiers and five reservists (who are perhaps best forgotten entirely).

Each battalion was accompanied by one 6pdr gun, manned by four gunners from each of the companies, but for wargames purposes the best way of representing this is to assign a three-gun 6pdr battery crewed by two figures to each regiment. In practice regiments were usually brigaded and the grenadier companies combined in a grenadier battalion, so on the table assign one grenadier battalion to every six musketeer battalions. Similarly, the guns could be amalgamated to give a six-gun battery crewed by four figures if desired.

The fusilier brigades each consisted of three battalions, organised as four companies of 165 men, giving a 20-figure battalion again.

The Foot Guard regiments consisted of three and two battalions respectively, each of six companies, so 30 figures per battalion on the table. Each of the jäger battalions had four companies as in a musketeer battalion.

In addition to the artillery accompanying the battalions, the Prussians had four foot artillery regiments, each of nine batteries equipped with six 12pdrs and two 6″ howitzers, plus the horse artillery regiment of 20 batteries, each with six 6pdrs and two 5.5″ howitzers. Obviously these would be concentrated where needed on the battlefield, but for wargames purposes a good ratio to use is one foot and two horse batteries to every 24 battalions, each represented by one model gun crewed by three figures. This does not sound much but with 24 battalions on the table there would be a further three six-gun 6pdr batteries in action with the battalions.

Cavalry regiments were generally organised into five squadrons, apart from two of the dragoon regiments which had ten. Each squadron consisted of roughly 160 men, giving five figures to a squadron. Exceptions are the hussar regiments, which all had ten squadrons, and the uhlans, who were 15 squadrons strong (the 'battalion' referred to earlier equalling five squadrons). Hussar and uhlan squadrons were weaker than the others, and four figures per squadron is appropriate on the table.

The Prussian cavalry was the best arm in the Army, but suffered in action through being spread all along the line rather than concentrated.

Post-1806

After Jena the King or Prussia was so

disgusted with his army that, with the exception of a few units which had performed well, he disbanded the lot and completely reconstructed it. In its new form it was much smaller, since Napoleon issued restrictions on its size, but had learned a severe lesson and became far superior in quality.

After the reconstruction there were 12 infantry regiments, each of one fusilier battalion and two musketeer battalions, all of four companies as before. These were organised in six brigades, each brigade containing two infantry regiments, one grenadier battalion, two regiments of heavy and one regiment of light cavalry. The exception was the Brandenburg Brigade which comprised the Guard infantry regiment, one battalion Leib grenadiers, one battalion Guard jägers, the Leib Infantry Regiment, Garde du Corps cavalry regiment, Leib uhlan regiment, one cuirassier regiment and two regiments of Leib hussars.

Little of Prussia's artillery survived the 1806 campaign, most of it being captured by France. However, it was gradually rebuilt with the loan of British guns to a total of 36 foot and nine horse artillery batteries. Each brigade received one battery of each type, the remainder being held back in Corps Reserves (a Corps being formed from two brigades).

Gradually Napoleon relaxed his restrictions on the size of the Prussian Army, especially as his thoughts turned more and more towards Russia, and by 1812 had allowed the formation of an extra nine cavalry regiments, two jäger battalions and a rifle battalion. All cavalry regiments were now of four squadrons, each of 120 men, so allow three or four figures per squadron–four is perhaps best.

At the same time the Prussians had devised a means of training more men for future armies by keeping a constant flow of conscripts into the field regiments and then back into civilian life as reserve troops. This enabled a large body of trained men to be built up ready for the moment when, through one reason or another, the restrictions were lifted.

The Prussian opportunity came, of course, after the invasion of Russia (in which they were also forced to contribute a contingent); mobilization of the reserves was, however, slow, and for the beginning of the 1813 campaign all the force the Prussians could muster was the regular army plus a number of Freicorps units-volunteer battalions organised in units of three or four musketeer companies and a jäger company.

By the beginning of the decisive autumn campaign, however, the reserve and Landwehr battalions were coming to arms in strength, with the result that the Prussians could muster no less than 90 line, 39 garrison, eight jäger and foreign and 151 Landwehr battalions, together with 247 squadrons of cavalry and 50 batteries of guns!

After the 1813-14 campaign, many of the Landwehr units were amalgamated into the Line, provincial grenadier battalions into the Guard, and a Guard rifleman battalion raised along with increased Guard cavalry. Not all of these saw action during the 1815 campaign however, but information on Blucher's army at Waterloo is plentifully available.

5 The British Army

Disciplined, well trained and led, the British Army caused Napoleon more problems in proportion to its size than that of any other nation, and played a particularly valuable rôle in pinning down veteran French units in Spain which could otherwise have been used in Russia in 1812.

In 1805 the British Army consisted of three regiments of Foot Guards, 96 regiments of foot (of which the fusilier and 'Royal' regiments should be treated as élite); two regiments of Life Guard cavalry, one of Royal Horse Guards, seven regiments of Dragoon Guards and 25 regiments of dragoons.

The normal regiment consisted of two battalions, one field and one depot (although many regiments had more), average strength of which was 700 men. The battalions were each divided into ten companies, average strength per company being 50-70 men. For wargames purposes therefore a battalion should be organised as 20 figures, two to a company. Two of the companies were called 'flank companies', the remainder were 'centre' companies. The

A battalion of 30 highlander figures drawn up in line to receive a French assault (Donald Featherstone and R. Tennant).

grenadier flank company served on the right of the battalion when drawn up in line, the light flank company on the left.

The Scottish regiments were usually stronger than the English line and should be represented by three figures to a company, 30 to a battalion.

Britain's experience in the American War of Independence a few years earlier, as well as filling the ranks with seasoned troops, led to some experimentation with light infantry. In 1803-4 Moore trained three regiments (43rd, 52nd and 95th) as light infantry, and in 1809 a further four were converted (68th, 71st, 85th and 90th). The degree of training was very high and these regiments emerged as the best light troops to see action throughout the Napoleonic Wars. They were organised identically to the line battalions.

The 95th had been issued with Baker rifles since 1800 and were well trained in their use. The 5th Battalion of the 60th Foot was similarly equipped, as were part of the light companies of the King's German Legion.

British cavalry was organised in regiments of four squadrons, although fewer than four were usually sent on campaign. Average squadron strength was 120-160 giving four figures per wargames squadron, say 12 to an average regiment in the field.

The artillery consisted mainly of 6pdrs with some 9pdrs and 5.5" howitzers; an average battery would consist of five 6pdrs and one 5.5" (horse); or five 9pdrs and one 5.5" (foot). Three crew figures per battery. There were also a number of Congreve Rocket troops, but their part in the three battles where they saw action was insignificant, and they are unlikely to be required on the wargames table.

Allied to the British were numerous foreign units, of which the best known and most significant was the King's German Legion, consisting of 15,000 Hanoverian troops who had escaped to England in 1803. By 1805 they consisted of four line and two light battalions of infantry, five artillery batteries, one dragoon and one hussar regiment (each of four squadrons). Organisation was along British lines. By 1806 their strength was up to eight battalions of line infantry, two of light as before, two dragoon regiments and three hussar regiments.

The organisation of the British Army did not change significantly throughout the wars, and unfortunately there is not room here to list all the units involved in the Peninsula and Waterloo campaigns etc, but these are readily obtainable in the many cheap paperback books on these subjects.

National characteristics

Every nation has its own characteristics influencing behaviour, mannerisms and attitudes; at its worst we see this in caricatures–the volatile Italian, the dour Swede, the brash American, the phlegmatic Briton, and so on. In a military force, these characteristics have a significant effect on the way a unit behaves under stress, and although one cannot be too precise, some attempt to allow for them should be embodied in our wargame rules.

For the sake of simplicity, two other criteria are studied at the same time: the type of training, discipline, and initiative practised; and the equipment with which an army is issued.

From a study of these factors we can arrive at a fairly simple evaluation of one troop type against another, which can be expressed numerically in the rules.

Taking the French first again, one can say that they were enthusiastic and untiring when on the offensive, but prone to discouragement in defeat. Proud of their achievements and, generally, confident in their leaders, the average French unit would carry out virtually any task asked of it under normal circumstances. However, discipline was often not as good as in other armies of the period, which could lead to units taking precipitate action.

To counteract this, the leadership from the Marshalls down to lowly NCOs was usually very good, and individual initiative encouraged. This meant that tactical decisions could be taken and acted upon very quickly, in the light of local experience, without guidance from above.

French equipment was on a par with most other armies, and although their muskets were superior to the Prussians' they were inferior to the British. French light infantry were better trained and more adept at their skirmishing role than any troops other than British light.

French heavy cavalry was poor in the early stages of wars, but after 1805-6, remounted on captured German horses, became superior to all other nationalities except the Russians. Their light cavalry tended to be a bit uncontrollable, but this was the same in all armies.

French artillery was exceptionally good, well equipped, trained and officered, and maintained a high rate of efficiency and morale throughout the wars.

The Austrians were generally the worst troops in Europe (other than the Spanish). They suffered from low morale, partly through the fact that the Austrian army was never a unified whole but consisted of an amalgam of Austrians, Hungarians, Serbs and Croats; and partly because the average Austrian soldier did not believe in the wars anyway.

They suffered from inadequate and outdated training as well as a lack of individual initiative or responsible leadership, with the result that they adhered to the old 18th Century linear tactics throughout the wars (see Tactics, page 40), manoeuvring and dressing ranks on the battlefield as if they were on the parade ground. This lack of training must be reflected on the wargames table: Austrian troops therefore take longer, in a wargame, to perform most evolutions than the French. Similarly, Austrian units should usually be made to attempt to deploy into line before entering a firefight or melée.

In the early days (ie, pre-1808) the Austrians also suffered from a lack of light infantry, and did not know how to use those they had in a skirmishing rôle. Thus Austrian units up to this date should normally be made to advance without benefit of a skirmisher screen.

Compared to the infantry, Austrian cavalry was very good, and was considerably superior to the French before 1806. After this it suffered through being mounted on the horses the French left behind, and never regained its battlefield supremacy. Austrian artillery was particularly bad initially although it improved after the reorganisation of 1806.

The Russian army was the strongest ever to face the French: strongest in manpower and strongest in morale. It too, however, suffered from the same deficiencies as the Austrians and Prussians, in that it was badly trained, led and equipped throughout most of the wars. The primary characteristic of the Russian soldier, stubbornness, was a great asset under these circumstances, and although one should not expect tactical miracles from Russian units on the tabletop, one should expect them to be extremely resistant under pressure, and very difficult to rout: this is reflected in their high morale ratings in the rules.

Like the Austrians, the Russians relied largely on outdated linear tactics and were very slow moving, especially since there was virtually no individual initiative allowed at the lower levels of command.

One tactical formation practised by the Russians which could be introduced though is the massive Divisional column, a solid body of men packed shoulder to shoulder like an ancient phalanx; the advantage of this was that, if menaced by cavalry, the outer ranks would simply face outwards while the remainder stayed facing the front—an 'instant' square in effect. It also makes a very good artillery target!

Russian cavalry and artillery were both very good and this is reflected in their high impact and fire factors in the rules. The cossacks, although tending to be rather uncontrollable, were also a useful force and very aggressive, albeit reluctant to tackle formed cavalry in the open.

The Prussian army was based on that of the Seven Years' War. Discipline was draconian and the average infantryman was more afraid of his officers than the enemy. However, once again leadership was poor, individual initiative virtually non-existent even in the higher command, and Prussian muskets were proverbially the worst in Europe. Linear tactics still prevailed and Prussian formations should generally be treated in the same way as Austrian.

However, Prussian morale remained high throughout the wars, initially through confidence in their own abilities and later through a desire for revenge.

Prussian cavalry, also like the Austrian, was very good initially but suffered later; by 1815 however it was almost as good as the French again. Prussian gunners were good but suffered through their poor organisation.

The British army of the Napoleonic wars was a good example of the exception that proves the rule. Basically similar to the Prussian in that discipline was very harsh and linear tactics were adhered to throughout, it nevertheless succeeded in beating the French at practically every turn. Why was this?

One reason is undoubtedly the steadfastness of the British infantry: ignorant gutter louts most of them may have been, they were certainly not afraid of the French and, like the Russians, were exceptionally stubborn when given a good defensive position.

Another is the superiority of their equipment and training: the Brown Bess musket was the best in Europe and fired a heavier ball, while being a small, professional, army, the men were well trained in their evolutions, could manoeuvre almost as fast as the French and could keep up an equal, if not better, rate of fire. This applied to the artillery, which was good, just as much as the infantry.

Where the British suffered was in their cavalry, which was well mounted but nowhere nearly as good as the French at any time and, moreover, was almost totally uncontrollable once unleashed: this too is reflected in the rules.

One primary factor which contributed significantly to British victories was the leadership; from men like Moore and Wellesley down to the lowliest subaltern it tended to be very good indeed—although there were exceptions—and when badly led British troops were no more invincible than the Prussians. To relect this British forces should be allocated a higher proportion of 'above average' officers.

Other nations: well, the Spanish were poorly led and equipped, as well as suffering very low morale generally and having a tendency to run away at the first volley. Treat them as Austrians but -1 on all fire, melée and morale factors etc. The Italians were not much better generally, and should be treated as Austrians. The German troops from countries such as Saxony, Bavaria and Württemberg were variable in quality but generally, being largely French equip-

French infantry battalion in 3-deep line with grenadiers nearest, fusiliers in the centre and voltigeurs on the right (Donald Featherstone and R. Tennant).

Three French officers confer on the road between Hal and Brussels during the Waterloo campaign (Frank Hinchliffe).

ped and trained, should rank as French but with a -1 on all factors. Poles were exceptionally good fighters and should count as Russians on all factors, as French for time taken to change formation. The Portuguese were generally fair troops and should be treated as British with a -2 on all factors, as should the Brunswickers and Dutch-Belgians. Swedes were good but reluctant fighters and are best treated as Prussians with a +1 on their factors for most purposes.

Obviously in a book of this length it is impossible to go into great detail on this subject, and for further information readers are referred to *Napoleon's Campaigns in Miniature*. However, I hope these guidelines prove useful.

Leadership factors

Leadership, that indefinable quality in an officer which causes his men to follow him anywhere, is, as we have seen, an important factor in warfare and some attempt should be made to utilise it in our wargames.

In the list which follows, selected Napoleonic generals and marshalls have been allocated two numerical factors for 'attack' and 'defense'. These factors are based on the individual's record of success or otherwise during the Napoleonic Wars, and while they must necessarily be somewhat subjective I feel they give a reasonable representation of each general's leadership capabilities.

The figures are used as follows: when assessing the morale of an advancing, attacking or charging unit, add to the total the leadership 'attack' factor for any commanding officer figure within 50 yards of the unit *and in sight.* Similarly, when assessing the morale of a standing or retiring unit, or when trying to halt a retreat or rout, add the leadership 'defense' factor for any commander figure within 50 yards and in sight.

Officer	**Attack factor**	**Defense factor**
French		
Napoleon	6	5
Bernadotte	2	3
Bessieres	3	3
Davout	4	5
Grouchy	3	4
Lannes	3	3
Marmont	3	4
Massena	4	4
Murat	5	3
Ney	4	4
Poniatowski	3	3
Soult	3	4
Austrian		
Alvintzi	2	3
Charles	3	3
Francis I	2	3
Hiller	2	3
Langeron	3	3
Mack	2	3
Melas	1	2
Schwarzenberg	2	3
Russian		
Bagration	3	4
Barclay	3	4
Bennigsen	3	2
Constantine	3	3
Doctorov	2	3
Kutuzov	4	5
Suvarov	4	4
Tormazov	2	2
Prussian		
Blucher	4	4
Brunswick	1	2
Bulow	2	3

One of the nicest personality figures made by any manufacturer is Hinchliffe's Eugene de Beauharnais, painted here in the uniform of the 13th Hussars.

Gneisenau	2	4
Hohenloe	3	2
Ruchel	2	2
Scharnhorst	2	3
Yorck	1	2
British		
Abercromby	3	4
Dalrymple	1	2
Moore	4	5
Picton	4	4
Ponsonby	3	3
Uxbridge	4	3
Wellesley	4	6

Obviously this list is far from exhaustive, and there may be other commanders you would like to add to it. The only way to do this is study their behaviour in the field and compare it with that of one of the generals for whom figures are given, and thus arrive at a comparative rating.

Similarly, many Napoleonic wargames are completely fictitious, and rather than having your forces commanded by a real historical figure, you may prefer to have fictitious generals. The easiest way to allocate leadership factors in these circumstances is to throw two 'average' dice and let the numbers thrown represent the factors.

Enthusiasts seeking even greater realism may also like to take into account the fact that even the best general was only human and could have 'off' days as well as 'good' days—Napoleon at Waterloo is the classic example of an 'off day. This could simply be represented by tossing a coin for each general: heads and it is a good day, so add +1 to his leadership factors, tails and it is an 'off' day, so deduct -1 from the leadership factors.

Table set up for a wargame with polystyrene ceiling tile hills, Airfix buildings, Riko hedges and trees, Bellona bridge and chalked in road and river. The French (nearest camera) are advancing to prevent a Russian crossing of the river. Note steel ruler and three dice in centre.

three

Into battle!

Having established our basic rules of scale and decided on the organisation and composition of our chosen army, we must now consider how these troops move and fight on the tabletop battlefield.

Infantry

The average infantry marching rate during the Napoleonic Wars was 60 paces a minute, or approximately 50 yards. This was for troops in a linear formation, when keeping the ranks straight was very important, and varied from army to army as we have already seen. March rates for the various troop types when in line are given in the rules for our 2½-minute Move.

Naturally, when troops were in column the need to keep ranks dressed correctly was neither so urgent nor so difficult, so a higher movement rate of 90 paces a minute, or approximately 75 yards, was average. Once again, however, the standard marching speed varied from army to army, and this is reflected in the rules. Further bonuses should be (and are) given for infantry charging into a melée.

Skirmishers and other light infantry acting independently with no need to maintain a formation of any sort were obviously able to move even faster, and 90 yards a minute would be average going. Note that this is *only* for light troops when in skirmish order, and that at other times they could move no faster than Line infantry.

Finally, it must be obvious that if the above rates are used for average movement across open fields etc, troops will be able to move faster when on good ground—especially roads; and slower when they have to hack their way through hedges, cross streams or marshy ground, or move through woods. Similarly, hill slopes will affect movement. Additions and deductions for special movement of this nature are given on page 54.

Having established how far our infantryman can move, the next question is, how often can he fire his musket? Very well trained infantry could get off as many as five aimed shots a minute, but this was exceptional and the average was nearer three. Furthermore, musket barrels very rapidly became clogged due to the coarse black powder used as a propellant, and this rate rapidly diminished. So as an average, two shots a minute would be about right, giving five

Close-up of the table shown on the left at the end of the first Move. Russian cossacks have charged over the bridge (top left) forcing the advancing French columns to form square. However, French lancers are advancing to meet them and the gun battery on the hill will inflict some casualties.

Above *over on the other side of the wargames table, the Russians are already advancing over the second bridge—jagers first, supported by cavalry and infantry columns—while French infantry and a horse gun wait for them on the ridge.* Below *on the right the Russians advance over the first bridge while their cossacks melée with the French lancers. Note the Old Guard in reserve by the foreground hill.*

shots in all during our 2½-minute Move. However, since a higher rate of fire would be usual for the first half-dozen volleys, we give our infantry a +2 on their fire effect for their first Move of firing.

Rifles were used by some units, notably British, but the breech-loader had yet to be invented and it was a dififcult job pushing the musket ball down the rifled barrel. Thus although rifles conferred a much greater range and degree of accuracy, they were slower to load than the conventional musket, and three shots per Move is all we can allow these units. The importance of deciding how many shots per Move a unit can fire will be seen later.

Apart from their fire function, infantry also had a rôle to play in hand-to-hand fighting, or melée as it is usually known. Since all men were armed approximately alike, with a bayonet on the end of their muskets, their capabilities in a melée must be similar. However, due allowance has to be made for the extra effectiveness of the heavier grenadier and Guards units, and this too is reflected in the rules.

A melée generally took place in two phases: the inital charge to contact; and the confused hand-to-hand scrum which followed. This means that we must allocate two different melée factors to all our troops, the first for their impact when they charge home, the second for their capabilities in a confused melée: this is really more vital in a cavalry than an infantry melée but still has to be catered for, hence the two sets of figures in the rules.

A number of other factors have to be taken into account when considering a melée. The first is the effect on the morale of a line of troops when they see a solid column of, say, French infantry bearing down on them, bayonets to the fore. It is a fact that wounds from bayonets were far outnumbered by wounds from musketry during this period, and this, coupled with accounts of actual encounters, serves to show that the melée was far less common than might be supposed.

Normally the defending unit, harassed by skirmisher fire and daunted by the mass of men rapidly approaching, would break and run before contact; or, if they stood firm, the line would pour such a withering fire into the head of the attacking column that it would falter and come to a halt, or retreat. So it is very important to assess the morale of both the attacking and defending units before a melée.

This is done as follows: when an assault is launched, the attacking troops are moved half way to their objective.

The defenders then fire their muskets at half effect (half a Move).

The attacking unit's morale must now be assessed. If it is good, then the attack will go home, and the defending unit must throw for morale. If they stand, then they may fire their remaining volleys at half way through the attacker's charge, causing further casualties on the attackers but not affecting their morale at this stage.

The first round of melée is then fought with the attackers using their impact factor *and* their confused melée factor, the defenders just their confused melée factor. If two units are charging each other, both obviously use their impact factors plus their confused melée factors.

One point which will be immediately apparent from a consideration of the above sequence of events is the need for the attacking unit to keep itself screened by skirmishers right up to the last moment unless it is to suffer very heavily from the defenders' fire.

Once the two units have made contact, casualties are worked out using the table on page 62. Whichever side loses the most men must retire (ie, is pushed back) 50 yards.

The losing unit now counts as unformed and must test its morale again. If it breaks the winners must take a control test to see whether they pursue, or stand and reform.

If the losing unit still stands, the winners—also now counting as unformed troops—must next take a morale test. If they break the losers may not pursue immediately but must reform.

If both sides stand then a second round of melee is fought using the confused melée factors, but the losers of the first round are disordered while their opponents are merely unformed, *and* count as troops 'following up'.

When one side or the other finally breaks, the winning unit must take a control test before taking further action. This test is to determine whether the unit is still under the control of its officers, whether it has lost discipline in the elation of victory, or whether it is shaken and similarly undisciplined.

Although the above procedure sounds complicated in writing, the stages will become clear if you practice moving a couple of infantry units through an advance-to-contact and melée, using the text as a guide to the sequence of events and the rules to determine casualties and morale.

One point ought to be made here on the subject of casualties. Basically, all the figures in the casualty tables refer to *men* killed, *not* model figures. The reason is simple when you consider that our model soldier represents 33 actual men: it is just unreasonable to assume that every volley of musket fire or round of melée will kill exactly 33 men, or multiples of 33.

Thus when calculating casualties, the procedure is as follows: for every 33 casualties, remove one model soldier from the unit under fire (thus 66 casualties, remove two figures). A record of 'remainders' should be kept on a battle record card, which is simply a sheet of paper with the names of all units in action written down one side.

To give an example: an infantry unit comes under fire and loses 45 men. Remove one figure (33 men) and jot down the number 12 beside the unit's name on the record card. On the next Move the unit comes under fire again and loses 54 men. 54 plus the 12 already 'carried' equals 66, so remove a further two figures from the table.

The amount of 'book keeping' involved in this method of determining casualties is only slightly more than that needed in simpler wargame rules and is crucial to the realism of the game.

A few final points on infantry: light troops are the only ones trained to act in skirmish order. Other troops must keep a recognised formation at all times or be considered unformed. Thus Line troops in houses or woods, or crossing rivers, hedges and ditches, are automatically unformed. Light troops in similar situations count as neither formed nor unformed for morale purposes, but as *formed* for melée purposes.

Light infantry, moving in skirmish order rather than close ranks, are obviously more difficult targets to hit, whether by artillery or musket fire. Thus in our rules a special deduction is made when the target is in skirmish order. Similarly, artillerymen always count as being in skirmish order when deployed (but not when moving). Cover, whether it be a ditch or the wall of a house, also confers a certain amount of protection, so troops behind cover suffer fewer casualties than those in the open.

Another fact that is obvious is the impossibility of launching a charge against troops inside buildings or behind walls, so all such melées are fought on confused melée factors only.

Similarly, all troops in the process of changing formation–for example, from column to line–are unformed until they are correctly aligned in the new formation, and must count as such if attacked.

Further anomalies and special circumstances will become apparent in battles you fight, and it is quite impossible to consider all of these. When resolving such situations, try to keep in mind what was likely to happen, what the reaction of the troops would have been in real life, and settle the problem amicably with your opponent. There is nothing more childish or infuriating than the wargamer who argues every tiny point to try to give himself an advantage, or who cites obscure instances of specific units which behaved in such-and-such a way at a particular battle in order to justify an action which would not normally be considered usual. If you feel this tendency in yourself, do try to curb it, or you will very rapidly find yourself without anybody to play with.

Cavalry

Moving on to the cavalry, I will not go into details as to how the various movement rates in the rules were arrived at since the procedure was the same as for the infantry. Anyway, it is obvious that light cavalry can move more quickly than heavy, that a charge Move when moving into contact will take place at a

Above *the Russians continue their remorseless advance on the left, the infantry columns screened by skirmishers while the cuirassiers engage their French counterparts in a melée.* Below *on the right the French lancers have defeated one squadron of cossacks and are now reforming to meet the Russian dragoons pouring over the bridge, while the second squadron of cossacks (on left) are engaging some French dragoons. French infantry have now also occupied the foreground cottage.*

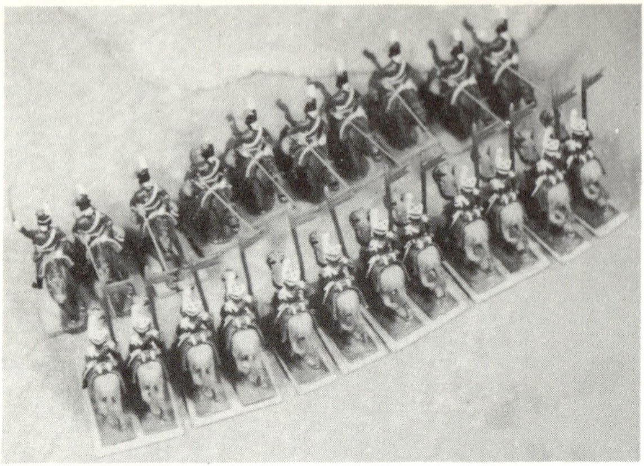

A typical cavalry melée between two of the most popular troop types in most wargamers' armies-Russian cossacks and French lancers. Note the bases are of different widths, representing the looser formation of the cossacks, which gives the French a one figure overlap on each end of the line.

more rapid rate than when a unit is simply moving from one point to another across the battlefield, and so forth.

Although most Napoleonic cavalry were armed with carbines, their function was not primarily a fire one. However, hussars and chasseurs à cheval were expected to be able to fire from the saddle as mounted skirmishers–surely a difficult task!–while dragoons, especially in the French army, were sometimes used in a dismounted rôle. And since it was possible, cavalry fire effect must be allowed for, and is in the rules.

The primary function of cavalry though was in the melée, and here we must break them down into three sections for consideration.

First were the 'heavies'—cuirassiers, carabiniers, Life Guards, grenadiers à cheval and heavy dragoons. These were big men, heavily armed and equipped, frequently with steel helmet and breastplate, riding big, heavy horses. Their function was that of a battering ram, to knock a hole in the enemy line through shock effect, and consequently their impact factor in a melée is high.

The cavalry in different armies varied in quality though and this is also reflected in the rules where, for example, French cuirassiers are worse than Austrian before 1808, better after (1808 has been taken as a convenient watershed since it was after this date that most Allied armies were reorganised, and also after this date that the French army began its slow decline in most arms).

The second cavalry type which deserves special consideration is the lancer. Most Napoleonic armies had a number of lancer units, although the French were rather late in acquiring them, and the British did not have any at all until after 1815.

Basically the lancer was an ordinary light cavalryman, and is treated as such for movement. It was his weapon which made him special, since the lance, in well-trained hands, was a devastating weapon as the British heavy dragoons found at Waterloo. In a charge the lance gave a light cavalryman roughly the same impact effect as a heavy cavalryman. This factor, coupled with his high movement rate, makes him one of the most devastating figures on a battlefield.

However, lancers should be used with caution for two reasons. First, if they are caught in a situation where they cannot *charge,* and thus use their lances effectively, they are no better in a melée than ordinary light cavalry. And secondly, although their impact factor is great, in the bounds of confused melée which follow, they are also no better than ordinary light cavalry.

The third type of cavalryman is the light horseman–the hussar, light dragoon and chasseur à cheval. His main function was reconnaissance before a battle and pursuit afterwards, although light cavalry could be and were thrown in to add numbers in a doubtful

Above *back on the left hand side of the table, the Russian infantry advance continues, while the jagers move to engage the horse battery. All the Russian and French cuirassiers are now engaged in the melée at right.* Below *a view of the right hand side of the battlefield from the Russian lines for a change, showing the Russian dragoons at the moment of impact with the French lancers. Note that neither side has had room to charge, putting the lancers at a disadvantage—but help is on the way.*

melée. As we have seen, they could also act as mounted skirmishers, advancing in extended order and firing their carbines from the saddle. This tactic is rarely practical on the wargames table, however.

Considering cavalry in general then, they have three main functions: to charge and counteract enemy cavalry; to menace infantry and force them into square (see tactics, page 41); and to pursue already fleeing troops. Heavy cavalry and lancers should be mainly used for the first duty, dragoons and other light cavalry for the other two functions.

A cavalry versus cavalry melée is fought in much the same way as an infantry melée, although in this case the chances are high that both sides will be charging to contact and so will use their impact factors in the first round. Neither side throws for morale before making contact *unless* they come under artillery or musket fire during the charge, since once unleashed they would normally plough ahead regardless, and it is only the distracting influence of heavy fire which could put them off.

The melée is then fought out normally, the only point worth making here being that heavy cavalry count extra against light cavalry (except charging lancers). Similarly, all cavalry count extra when charging infantry–see page 61.

After the first round of melée, whichever side has lost the greatest number of men is pushed back 100 yards and must test morale; and so on until one side or the other breaks. The winner must then take a control test, this being particularly important for cavalry who were frequently very impetuous and likely to go haring off towards the nearest enemy unit, completely out of control (the British were worst in this respect).

Artillery

Artillery, as discussed briefly under 'organisation', came in three types: horse guns, field guns, and howitzers. The first category consisted of the light 3,4 and 6pdrs, the second of 6,7,8,9 and 12pdrs, and the third of 5.5″, 6″ and 8″ plus the Russian licornes. All guns were capable of firing ball and canister, only howitzers could fire shell (see footnote, page 42). As we have already discussed, the guns were divided into batteries, which on the wargames table are represented by a single model gun mounted on a triangular base.

Artillery ranges and effectiveness varied according to the calibre of the gun, and these are all reflected in the rules, which have been averaged to give the normal casualties a battery could be expected to cause at given ranges and using the different types of ammunition. One point worth discussing however is that of penetration.

Obviously, when a cannon ball hit the ground, it did not stop dead but continued bouncing and rolling for a considerable distance. Even when almost at a standstill, the ball was a considerable menace, as many an unfortunate soldier found by trying to stop the passage of a ball with his foot! In fact, a special form of fire called ricochet fire was often employed, where instead of aiming for the centre of a formation of enemy troops, the gunners would aim just in front of them and trust to the ball continuing bouncing through the formation. Naturally this tactic could only work when the ground was fairly firm–it was useless after rain–and was not nearly so effective if the ball was bouncing uphill. Thus when considering the penetration figures on page 57 of the rules, *halve* all distances after one Move of heavy rain, two Moves of light rain, or when the target is on a facing hill slope.

The next question with regard to artillery (and musket) fire is, what proportion of fire can be loosed against a target which is only within the battery's field of fire for part of a Move? And similarly, at what range does one calculate fire if a target is at one distance at the beginning of its Move and another at the end? Both of these questions can be resolved once we know how many shots a gun can fire during the course of a Move.

We have already seen that a musket can be fired five times during a Move, a rifle three. But how many shots can an artillery battery fire?

This depended largely on type, lighter guns being far quicker to reload than heavier ones. Howitzers of all calibres

The next three moves see some dramatic events on the left. Top *the Russian infantry reach the ridge and are met by a volley of French musket fire, while on the right the cavalry melee continues, neither side having secured an advantage.* Centre *the first Russian infantry battalion retreats unformed as a result of an adverse morale test following the French musket fire. However, the second battalion has charged home. At the same time the French horse gun, harassed by the jagers, has limbered up and is moving to a new position, while on the right the Russian cuirassiers have at last pushed the French back.* Bottom *the French infantry line is being pushed back. the French cuirassiers are in full retreat, while the French grenadiers on the right hand ridge have formed square.*

were slower to reload than guns in general, and count as heavies in this instance. Working from contemporary figures, we arrive at the following list: a 3 or 4pdr could loose off eight rounds in our 2½-minute Move; a 6 or 7pdr six rounds; an 8 or 9pdr five; and a 12pdr or howitzer, three.

This gives us a proportionate fire effect chart as follows: the top line shows the amount of Move during which the target is in view, or, the amount of Move left for firing after a battery has changed position. The vertical column on the left shows the type of weapon firing. And the figures in the chart itself show what proportion of its fire the battery can bring to bear in the time available.

leaving approximately 1/8 of a Move in which the cavalry are in sight and the guns can fire. Looking at the top row of the chart for 1/8 and then down for a 12pdr, we find that the battery cannot fire at all in the time available, so the cavalry escape scot-free.

One final example: a unit of infantry are charged by other infantry. At the beginning of their Move the attackers are out of musket range, and only enter range after 1/5 of their Move has been taken up. Thus when the defenders loose off their volley at half-way through the attackers' charge, they have already 'lost' one possible volley and their fire must be taken at half of 4/5 effect, ie, 2/5. The infantry then complete their charge and the defenders fire again at

	1/10	1/8	1/6	1/5	1/4	3/10	1/3	3/8	2/5	1/2	3/5	5/8	2/3	7/10	3/4	4/5	5/6	7/8	9/10
3-4 pdr	O	1/8	1/8	1/8	2/8	2/8	2/8	3/8	3/8	4/8	4/8	5/8	5/8	5/8	6/8	6/8	6/8	7/8	7/8
6-7 pdr	O	O	1/6	1/6	1/6	1/6	2/6	2/6	2/6	3/6	3/6	3/6	4/6	4/6	4/6	4/6	5/6	5/6	5/6
8-9 pdr	O	O	O	1/5	1/5	1/5	1/5	1/5	2/5	2/5	3/5	3/5	3/5	3/5	3/5	4/5	4/5	4/5	4/5
12 pdr & How.	O	O	O	O	O	O	1/3	1/3	1/3	1/3	1/3	1/3	2/3	2/3	2/3	2/3	2/3	2/3	2/3
Musket	O	O	O	1/5	1/5	1/5	1/5	1/5	2/5	2/5	3/5	3/5	3/5	3/5	3/5	4/5	4/5	4/5	4/5
Rifle	O	O	O	O	O	O	1/3	1/3	1/3	1/3	1/3	1/3	2/3	2/3	2/3	2/3	2/3	2/3	2/3
Mounted Carbine	O	O	O	O	O	O	O	O	O	1/2	1/2	1/2	1/2	1/2	1/2	1/2	1/2	1/2	1/2
Carbine	O	O	O	O	1/4	1/4	1/4	1/4	1/4	2/4	2/4	2/4	2/4	2/4	3/4	3/4	3/4	3/4	3/4

Since this is a little complicated, let us examine a couple of examples. First, a horse artillery battery which has moved 60 yards–say 1/5 of its Move; it then takes a further 1/5 move to unlimber, leaving it 3/5 of its Move in which to fire. Looking along the top row for 3/5 and checking this against the line for a 3-4pdr, we find the battery can loose off 4/8 or half its normal full fire.

Second, a 12pdr battery across whose front is moving a light cavalry squadron. If, say, the cavalry occupied 120 yards of their Move before coming within the battery's field of fire, this is roughly 1/3. They then move across the battery's field of fire a further 120 yards before completing their Move outside its field. Thus they are a *possible* target for 1/3 of their Move. Now it takes the battery gunners 1/5 of their Move to re-align their guns on the new target,

quarter distance, but since 1/4 is greater than 1/5 they can only let fly 1/5 volley at the new range.

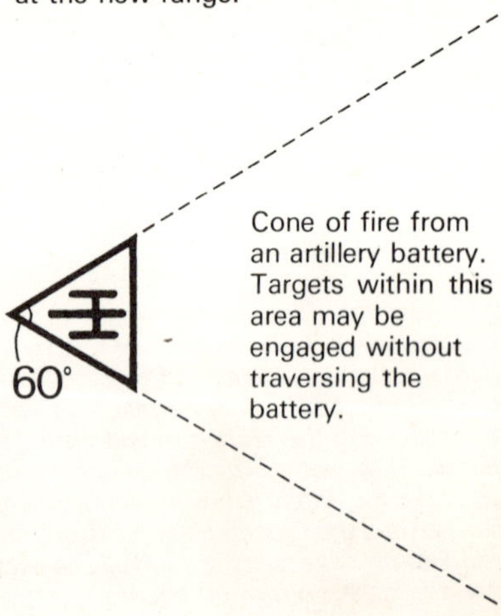

Cone of fire from an artillery battery. Targets within this area may be engaged without traversing the battery.

Back on the right, things are going better for the French. Above *the grenadiers à cheval have joined in the melée with the Russian dragoons and the French infantry are rapidly advancing, while the remaining cossacks are being chased off by lancers and French dragoons.* Below *the Russian dragoons have now been chased back across the bridge, leaving the Russian infantry on this side of the river completely isolated; they cannot even retreat without having French cavalry on their backs.*

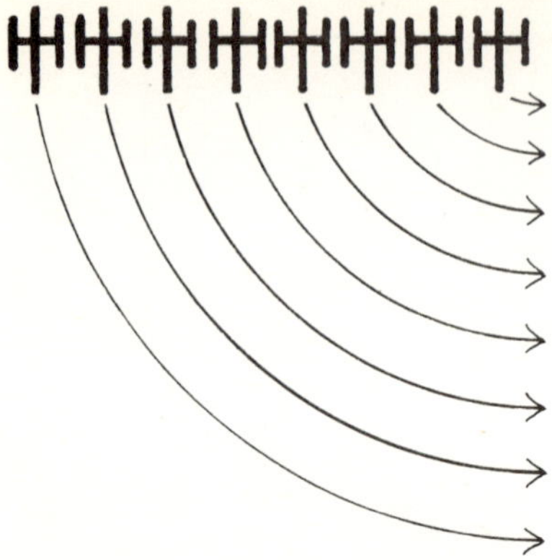

Diagram to show why it takes longer to traverse a gun battery through 90° than 180°; in the latter case the gunners simply have to turn the guns back to front, but in the former they have to move as shown here—a considerably greater distance.

I hope these three examples serve to show you how to use the chart; if you are in any doubt in a given situation it can usually be resolved by going back and moving again in smaller stages, say a quarter of a Move at a time.

Morale

In the preceding pages I have frequently mentioned morale, but have not gone into great detail. Morale is that indefinable characteristic which causes one unit to turn and run, the next to stand and fight, despite essentially similar circumstances.

Morale can be affected by a wide variety of circumstances: discipline, training, esprit de corps, leadership, position, formation, casualties, fatigue, hunger and a host of others. In the morale chart in the rules the most significant of these are expressed as numerical factors which are added to or deducted from a unit's basic morale factor, as laid down in the national characteristics tables.

To assess a unit's morale, first take its basic morale factor, then throw two average dice and add the *difference* (eg, if you throw a 3 and a 5, add 2; 2 and a 5 add 3 etc). Keeping this new total in mind look down the column under attack (a) or defense (b), depending on whether the unit is attacking or advancing (a), or standing, retiring, retreating or routing (b).

Add or deduct all figures which apply to the unit's circumstances—for example, has it lost any casualties? Is it formed? Is it threatened? Has it got secure flanks? Has it won or lost a melee? Is it under fire?—not forgetting to add the leadership factor of any general officer within 50 yards and in view.

The total—the higher the better—represents the unit's morale state at that particular time in the battle, and should then be compared with the morale test results table which decides the unit's further action.

Generally speaking, the morale rules have been designed so that units will break fairly easily unless they are in strong, well-supported positions, have few casualties and are well led, which reduces the number of melées as in real life.

When looking at this period, it is worth remembering that 30 per cent casualties from a battle was considered very high indeed, and most encounters were decided before either side had lost anything approaching this percentage. There *were* bloodbaths—Borodino for example—but generally an army's lines would begin to crumble and its men run before casualties reached a high level. Pundits will point to the action of the 957th or whoever at some obscure skirmish in Outer Mongolia and show how they fought to the last man, but such incidents were the exceptions which proved the general rule: most units are finished as fighting forces before they are reduced to half-strength.

As a corollary to this, you will also find that units which have broken due to adverse circumstances rather than high casualties will rally quite easily, and can be brought back into the battle fairly promptly.

This ends our assessment of battlefield situations. Further points which may not be immediately clear are explained in the rules, which should be read with care. However, the only way to become really fluent in wargame tactics is not by reading, but by playing, and the sooner you get the troops out and start fighting the sooner all the pieces will click into place.

Above *on the left things look really bad for the French. Their cuirassiers have routed completely off the table, taking the horse battery with them as a result of a morale test, while the Russian cuirassiers are left in possession of the field. The Russian infantry have now pushed the French off the ridge, and it looks as though the French infantry will soon disappear too. The second Russian infantry battalion has also reformed and is advancing again in the background. However, perhaps all is not yet lost.* Below *the Russian infantry around the other bridge are pinned in square by two French cavalry figures, releasing the rest of the lancers and grenadiers to go to the aid of the left flank. The French infantry, including the Old Guard, are rapidly moving up, when the extra firepower from their line will soon decimate the Russian square. So, although that is where we must leave it, it is still very much anybody's game.*

four

Napoleonic strategy and tactics

Strategy is basically the art (or science) of moving armies across country in such a way as to bring them together at a chosen battlefield, outflank an enemy position or take a particular objective. It is, essentially, the art of war in large. Tactics is the art of the battlefield itself, of where to position troops at the beginning of an encounter, where to attack, where to defend, and in what formations.

Strategy will not concern most wargamers until they begin to play a campaign, but a brief assessment of the main points is still useful.

Most European armies of this period were very slow moving, being tied to magazines and depôts with their ponderous wagon trains for supplies. The French army on the other hand relied to a large extent on the food and forage to be obtained in the countryside through which it passed, and was thus much more mobile.

Similarly, most European nations moved their armies in self-contained units, each one of which could occupy several miles of road. This meant that the rear end of the column could just be leaving its camp from the previous night at the same time as the head of the column was setting up its bivouacs for the following night! Once again, this contributed to the slow movement rate.

Because of their dependence on local supplies, the French could not afford to move whole armies together, and so devised the Corps system (which was eventually adopted by the other nations). Under this system an army would be divided into three or more Corps, each consisting of a balanced force of infantry, cavalry and artillery, and moving on a common axis although not along the same road. At the beginning of an offensive these Corps would be widely dispersed over a very long front, but would gradually converge towards the point selected by Napoleon for a decisive encounter. Thus by the

French attack in column against a British line on a ridge—a foolhardy undertaking unless the attack is made in overwhelming strength. Note fire dice.

Two French battalions advance in l'ordre mixte with grenadier companies in the centre and voltigeurs in a skirmish line in front.

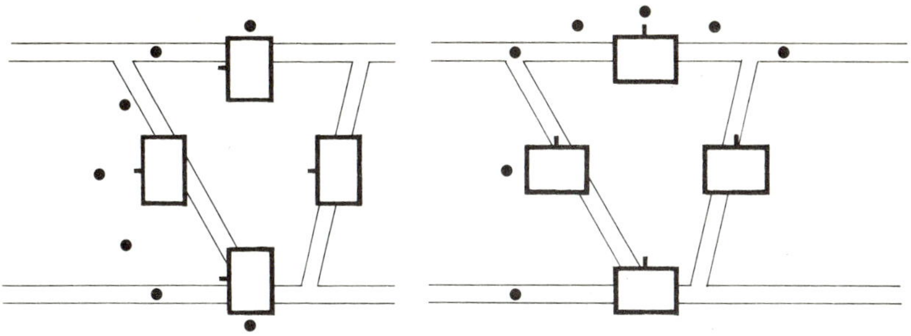

Four French Corps on the move, showing the flexibility of the 'Battalion Carre' system. This diagram clearly shows how the axis of advance can easily be changed to either flank or rear, the cavalry scouting screen similarly falling in line with the new direction.

time action was imminent the various Corps would all be within two, or at the most three, days' march of each other, so that if one was menaced by a superior force reinforcements could rapidly be rushed to the spot. Each Corps was fully independent and was designed to be able to hold out against odds of two or three to one for a sufficient length of time to enable reinforcements to arrive, so that the unwary adversary could find himself trapped into an escalating battle of attrition against a continually growing French force.

This Corps system also allowed great flexibility in so far as the army's line of march was concerned, since any Corps could act on a moment's notice as advance guard, rear guard, or flank. The accompanying diagram explains this.

The main aim of strategy is always to bring a greater number of men and guns together in one place than the enemy can muster–to 'be there fustest with the mostest' as more than one general is credited with saying. There were various ways of achieving this. One was simply to move so quickly that the enemy was practically mesmerised into inaction, then outflank him and defeat him as Napoleon did to Mack at Ulm.

Another, especially when faced with two or more opposing armies (as Napoleon was throughout most of the wars) was to detach a relatively small force to pin down one of the enemy armies in a delaying action, and concentrate the rest of the army against the second foe. Having defeated him, the main body could then swing against the first enemy army. This ploy, called the 'strategy of the central position', is illustrated in the accompanying diagram, and was used most successfully by Napoleon at Jena-Auerstädt (where the French pinning force, under Davout, actually succeeded in defeating a much larger Prussian army at the same time as the main French army was giving the main Prussian body a hammering at Jena).

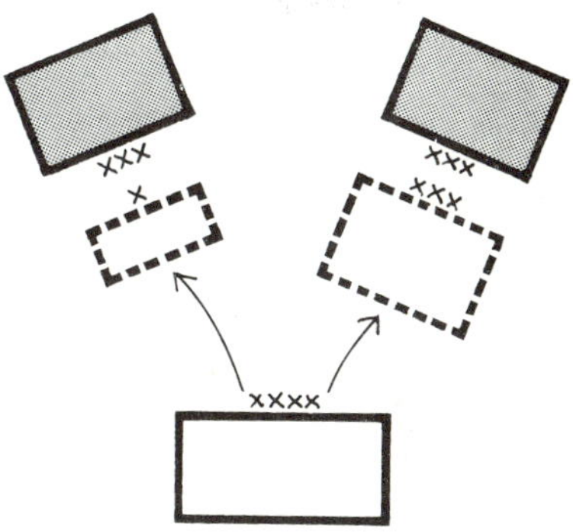

Strategy can also involve luring your opponent into a trap, as Napoleon did before Austerlitz by pretending to withdraw, thus persuading the Allies that he was afraid of battle, then encouraging them to attack at a selected point (in this case his right flank) by making it appear weak so that the main French attack through the centre could split the Allied army in two.

All these factors can be incorporated into a wargames campaign, but there is not room here to go into Napoleon's strategy in greater detail. A complete analysis of Napoleon's art of war is given in David Chandler's superb book *The Campaigns of Napoleon,* which every serious student should read. Another useful though brief reference is *Strategy & Tactics* magazine No 32, 'Napoleon at war'.

Tactics is the art of manoeuvre on the battlefield and is largely dependent on two factors: the mobility and manoeuvrability of the troops, and the range and destructiveness of the weapons.

In Napoleonic times the basic movement rate was that of the infantryman slogging his way through mud and undergrowth while at the same time trying to maintain some sort of formation. The second restricting factor was the range of the musket, which was only effective up to 200 yards and rarely fired even then until the enemy were within a hundred yards.

In the 18th Century the usual tactical formation was a line of men drawn up in three or more ranks; each rank would fire a volley in turn then reload while the other ranks were firing. Controlled volley fire of this sort, from disciplined troops, was far more effective than is usually imagined, and it was this basic tactic which the armies of Prussia, Austria, Russia and Britain employed during the Napoleonic Wars.

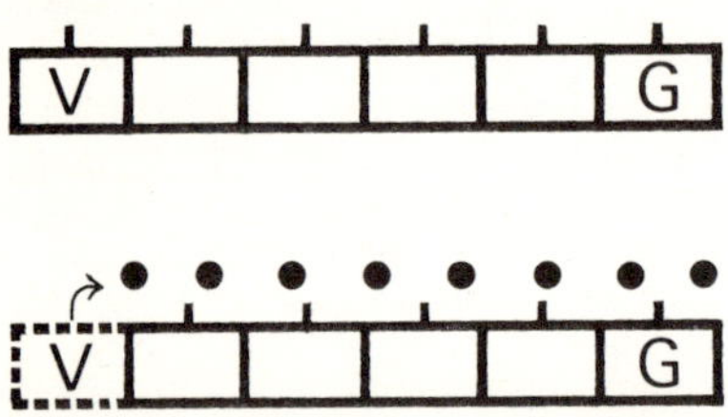

French battalion in line with voltigeur company on left and grenadiers on right. In the lower diagram the voltigeur company is deploying into skirmish order.

The French army, on the other hand, driven by necessity in the early stages at least partly due to the poor quality of the troops in the Revolutionary armies, had evolved some completely new tactics which worked well against all except the most steadfast foe. Basically the new tactics involved the use of small, highly mobile columns of men, who relied on the shock effect of their impact on an enemy line to break it rather than fire effect.

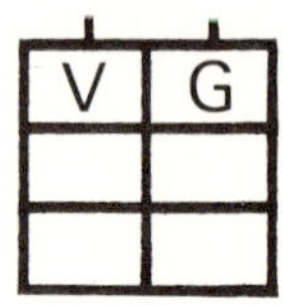

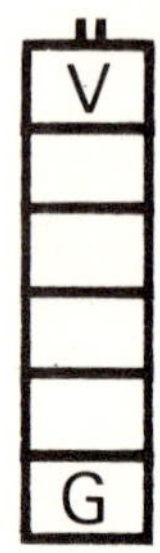

Two variations on a battalion column of six companies. V voltigeurs, G grenadiers.

This was also essential because French musketry training was abyssmal, and in fact whole battalions were sometimes sent into action without any ammunition anyway!

Obviously a dense column of men was very vulnerable to close-range volley fire, so a further innovation in the shape of swarms of skirmishers was introduced. These skirmishers' task was to fan out in a cloud in front of the advancing columns and pour shot after shot into the ranks of the enemy force, weakening the effectiveness of their volleys and frequently demoralising them before they even fired a shot in return. These tactics were particularly effective against an army like the Prussian, whose troops would frequently still be manoeuvering into line with all the correct intervals, for all the world as if they were on parade, while the skirmishers harassed them with a continual fusilade of fire. The French skirmishers would continue their diversionary work against the enemy line until practically the moment before the advancing column impacted, preventing the enemy ever getting in a decisive volley.

This tactic was obviously less effective when the opposing troops were on higher ground, since they could fire their volleys over the heads of the skirmishers into the columns behind (as happened so often in Spain).

It must not be imagined, however, that all French attacks were launched in column. Frequently, especially when facing an opponent in a strong defensive position, they would deploy, or attempt to deploy, into line before pressing the attack home. And in defense, of course, the French used line as much as anybody else.

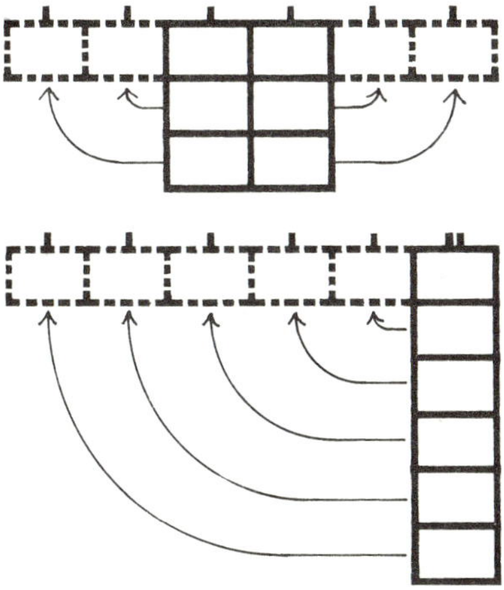

Deployment of companies from column to line. When going from line to column procedure is the same in reverse.

A variation on the column attack which was, in fact, the most successful infantry formation devised by the French, was *l'ordre mixte,* whereby three or more battalions would advance in a combination of line and column (see diagram), thus bringing to bear both the

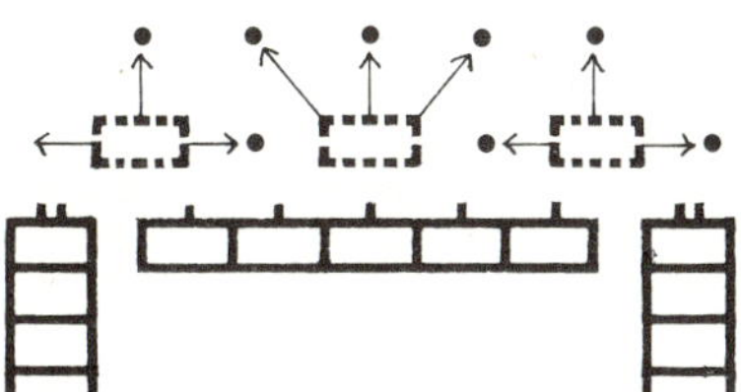

L'ordre mixte

fire effect of a linear formation and the shock effect of a column.

When menaced by cavalry, there was only one thing infantry could do, and that was form a square as shown in the accompanying diagram. The square

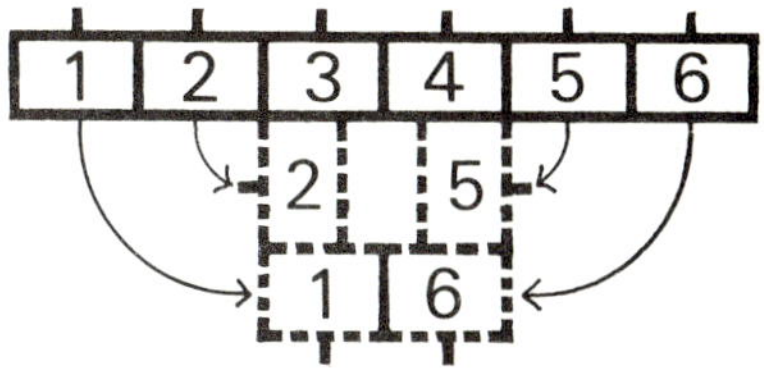

Above *forming square from line.* Below *forming square from column.*

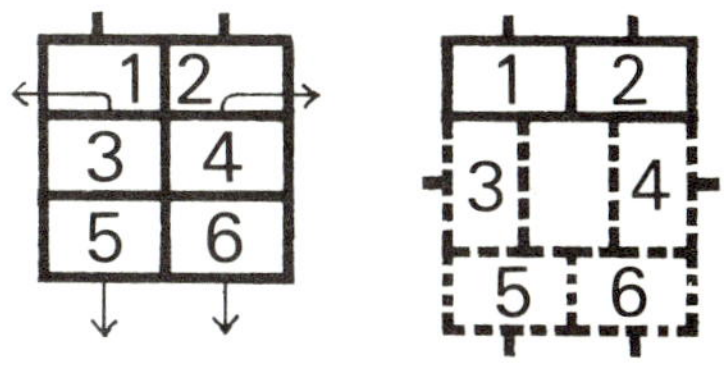

formation had no flanks the cavalry could sweep round, and so bristled with bayonets that no horse would charge home. Squares were usually formed up corner to corner as shown so that the

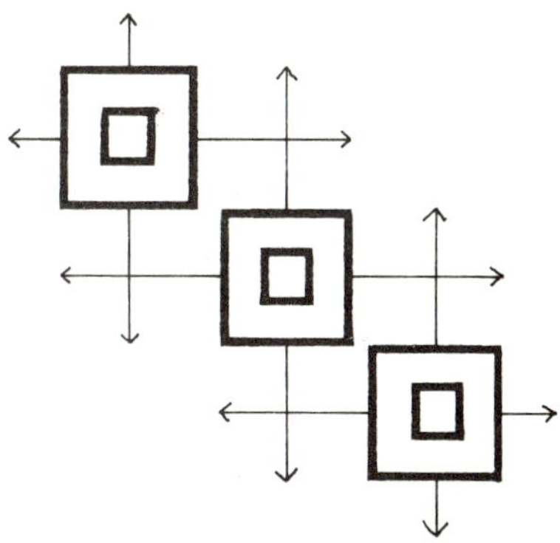

maximum number of muskets could be brought to fire on the cavalry milling around them. There *are* instances of infantry squares being broken by cavalry alone, but they were few and far between so in our wargames rules a formed square cannot be charged by cavalry. A 'shaken' square can though.

However, a square had one great disadvantage: it was virtually immobile, and for wargames purposes we shall consider it completely immobile. This meant that a general who thought ahead would have his cavalry attack supported by horse artillery, which could fire point-blank into the densely packed infantry squares while the cavalry hovered nearby waiting their chance to break in. Once cavalry were inside a square all hell was let loose and the infantryman who survived was extremely lucky.

An alternative to attacking the squares with cavalry once the horse artillery had done its work was to bring up a column of infantry to charge in: had Ney brought artillery and infantry support in behind his massed cavalry charges at Waterloo, Wellington's redcoats would have been massacred—and they usually are in a wargame!

Other than forcing infantry into square, the other cavalry use was, of course, against the enemy's cavalry. There was little tactical finesse here, the sole aim being to get more and heavier men into contact and beat your opponent down by sheer weight and numbers. This particularly applied to the heavy cavalry, since the light—hussars and chasseurs à cheval—were generally reserved for pursuit duties once a foe was on the run.

Artillery tactics deserve a book on their own, and certainly in wargames the guns are more frequently mishandled than any other arm. Basically, the heavy batteries were sited at the beginning of the battle, frequently massed together into Grand Batteries to pour a concentrated fire on a selected point in the enemy line.

Their main target was the enemy artillery, since an infantry assault against concerted fire from enemy guns was almost certainly doomed to failure. On the tabletop therefore you should select your point of attack and concentrate your artillery fire against any enemy guns on that front. Of course, if guns were menaced—for example by cavalry—they would change target quickly enough, and you should beware of concentrating on one target to the exclusion of others which may be more attractive—fire down the flank of an enemy column or line is particularly devastating.

The purpose of the horse artillery was to get in as close to the enemy as possible and bombard him with canister fire,* moving speedily out of the way if threatened and firing against another target. As we have seen, if horse artillery is sent in in support of a cavalry attack it can be especially effective.

In wargames terms there are a number of points which can be made about tactics. When you first examine the battlefield, assess what you know about your opponent and the troops at his disposal. Try to work out where he is most likely to launch his main attack. Where is he going to site his guns? Where is he going to mass his cavalry?

---

* There were four main types of artillery ammunition in use during this period. The most widely used was Ball—large heavy balls of iron which would plough through an infantry formation knocking men down like skittles, and which would continue bouncing for a considerable distance on dry ground. The second main type was Canister, a development of the earlier Grapeshot in which a thin metal case would be filled with musket balls. These would spread out in a cone of fire when discharged, and were quite devastating at close range. The third type of ammunition used was Shell, which could only be fired from howitzers, and consisted of a hollow metal ball packed with an explosive charge and, sometimes, musket balls as well. The last type of ammunition was called Shrapnel, and was only used by the British during this period. If a shell burst above the ground in the middle of an enemy formation it could be very destructive, but it was difficult to time the fuses correctly and shells could be kicked away to explode harmlessly in a ditch or on the other side of a wall. shell was also used to set fire to wooden or thatched buildings, forcing the defenders into the open. The ranges and differing degrees of effectiveness of these four types of ammunition are all allowed for in the rules.

Then evaluate your own possibilities. Are there any natural defensive lines on your side of the table? Are there any likely avenues of attack (for example,

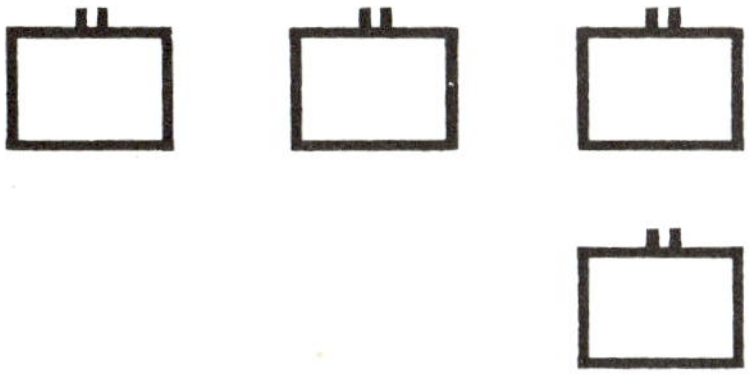

Two typical Napoleonic attack formations. Above *battalion advance in line, right flank reinforced.* Below *battalion advance in echelon, right flank refused.*

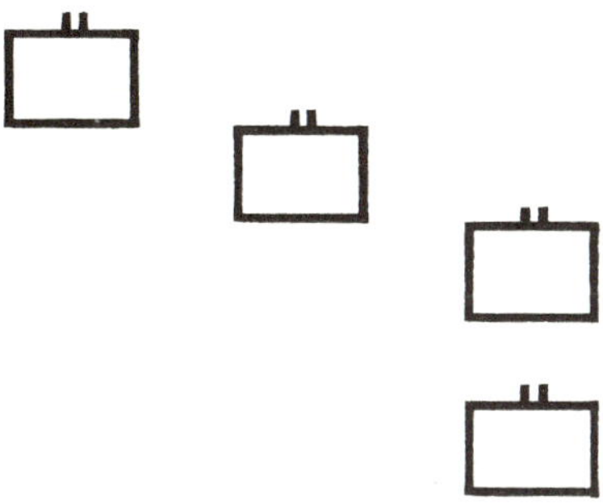

through woods) where your opponent will have difficulty in counter-attacking or bringing down effective artillery fire? Where are your cavalry most likely to be effective–it is no good concentrating them at a point on the table where they cannot manoeuvre. And where are you going to site your guns so that they can both hit the enemy artillery hard and support your main assault?

Frequently it is best to assume a defensive posture at the beginning of the battle, refusing to commit your troops to an attack until you have seen your opponent's dispositions. Let him attack you, then, when you have repulsed him, follow up with a counter-attack of your own. Sometimes this is impossible–the terrain or the troops at your disposal will not permit it–but certainly when you can achieve it a strong defense followed by a well-timed attack is the quickest and surest way of winning. Unfortunately, having also read this book, your opponent will be trying the same!

Generally, it is best to try an attack on one or other of your opponent's flanks. If this succeeds it will break him at best, force him to form a new flank at right angles to his existing front at worst, and tie down his reserves in any case.

Attacks should always be mounted with all the force you can possibly muster, including artillery support. If you can rout one unit it will shake all the surrounding units and a chain-reaction can set in with the units around it retreating and retiring all over the place, leaving a magnificent gap into which to pour your cavalry.

So, if you can, concentrate on breaking one unit on your opponent's flank–the results can be staggering! Similarly, in defence try to protect your flank units by giving them cover or the reverse slopes of hill to hold, thus preventing the same thing happening to you.

Other than this, battlefield tactics is largely a matter of common sense. Try to position your guns with a wide field of fire, preferably on high ground, and if possible with overlapping fields of fire so they can support each other. Avoid positioning units in such a way that your opponent can bring fire on their flanks–especially cavalry.

Above all, keep moving all the time. Seize the initiative at the earliest possible moment and keep hammering away so as to keep your opponent off balance. But do think first: don't be rushed into precipitate action. Position your forces carefully at the outset and work out exactly what each is supposed to be doing before moving it.

Don't be panicked into rushing reserves into the fray before it becomes absolutely necessary, but by the same token don't be afraid of pouring troops into an attack or counter-attack when the circumstances are promising. Bear in mind at all times the capabilities, strengths and weaknesses of the troops under your command, and exploit these to their full. Above all, don't be too hesistant: battles are lost more often by cautious commanders who refuse to commit their troops than by bold ones who commit everything.

five

Organising a wargames campaign

Campaigns are to battles as bread is to butter: on its own the latter soon palls. What does it matter if you win a battle? At the end of the day the players pack up their troops and go home, and it is all over until your next meeting. The obvious way to get round this is to organise a game league table in your local club, which is beyond the scope of this book, or alternatively and preferably a campaign.

A campaign can be as simple or as complex as you wish, ranging from a simple action against a neighbouring country with victory or defeat hanging on a single battle, to a complete re-enactment of the entire Napoleonic Wars, taking in every facet of organisation, supply, diplomacy, weather, recruitment, sieges, tactical and technical invention and so on. This would be too complicated to enter into here, so I shall restrict myself to some suggestions for organising a campaign between four or more players which could take up several months of real time and involve some exciting battles.

The first requisite for a campaign is an umpire - someone who is prepared to collate the movement orders issued by the opposing commanders, decide when contacts have occurred, arbitrate on the rules and generally 'play God!'

The second requirement is a set of reasonably large-scale maps of the terrain to be fought over, one set for each player and one for the umpire. Large-scale physical maps of Europe are difficult to come by and expensive, so most campaigns end up being fought over 'fictitious' terrain.

The most popular country over which to fight a campaign is Ireland, since large-scale (4 miles to the inch) maps are cheap and easy to come by; the country includes a wide variety of terrain types, from open river plains to heavily wooded hills, mountains, lakes, cities and rocky escarpments. Moreover, the towns are mainly small, the roads undeveloped, and the country lends itself to division into a number of wargames 'countries'. I would suggest that the northern section (basically Ulster extended slightly to the west) should represent France, the Southern portion Russia, while the central portion can be

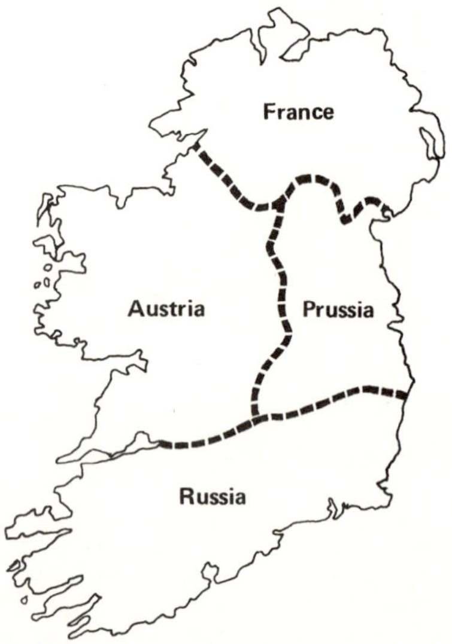

divided north-south into 'Prussia' and 'Austria'. Of course this is only a suggestion: a campaign on a smaller scale can be fought just as easily by splitting the Isle of Wight in two, or taking two or more adjoining British counties and giving each the name of a country. Alternatively, maps of the actual campaign areas could be purchased at a price, or a completely fictitious terrain drawn up. It really doesn't matter.

Having got a map the next thing to decide is what period you want your campaign to take place in, since as we have already seen the organisation and composition of the Napoleonic armies changed considerably over the years.

1805, 1812 and 1813-14 are all popular, but again it is really up to you.

Having decided your period, the umpire should now set about drawing up lists of the troops involved. To a large extent this should be based on the number of wargames figures available to you, but not exclusively since the armies should be split into a number of Corps or Divisions and are unlikely to ever appear on the same battlefield at the same time. Ideally the strategic organisation should parallel as closely as possible that used by your country in the chosen period, and certainly the tactical formations should be historically accurate.

Having come this far, the next question is: how many miles can your troops march per day? This can be resolved by reference to accounts of real campaigns, but for simplicity's sake I have set out average marching rates for the five major powers in the table below.

Given this, each player can now work out how far each of his units can march from their starting positions per day or week; the orders for each unit are then submitted to the umpire, who works out the positions of each player's troops. When two forces come into contact the players are told and a battlefield is set up corresponding as closely as possible to that revealed on the map.

One factor to remember is that armies did not move in one big lump, but were stretched out, frequently over several miles of road. Thus when a contact takes place, unless it occurs late in the day which would enable the armies to dispose themselves in line, units will arrive on the table one by one. It is thus essential that all players keep lists showing their units' order of march so that they can be brought onto the table in the correct sequence. Note that in this period the infantry usually marched at the sides of the road, with cavalry and artillery occupying the centre.

Using cavalry as scouts and screens can be quite profitable, since they can enable you to discover an enemy's position and line of march (while ideally concealing your own), and thus enable you to lead him into a strategic trap where he either has to fight you at unfavourable odds, or you can outflank him.

One perplexing point in setting up a battle is the relationship between time scale on the map and time scale on the table. The best way to resolve this is to rule that one Move on the table equals one hour on the map, so that a French unit six miles away from the battlefield could reach the scene of action in three game moves if on a road or six game moves if forced to march cross-country. Force marching would, of course, halve these times, but troops who have force-marched for three hours or more should count as tired when they arrive on the table.

Of course, while all this is going on, it is essential that you keep in touch with your own units. If the campaign is a simple one with only one player in charge of each country, there is no problem—

Average daily and hourly marching rates

Nationality	NORMAL				FORCED			
	Road		Cross-country		Road		Cross-country	
	P/hr	P/day	P/hr	P/day	P/hr	P/day	P/hr	P/day
French	2	16	1	8	3	24	1½	12
British	1¾	14	⅞	7	2¼	18	1⅛	9
Russian	1½	12	¾	6	2	16	1	8
Austrian	1¼	10	⅝	5	1½	12	¾	6
Prussian	1	8	½	4	1¼	10	⅝	5

you always know where everyone is. But if each side involves two or three Corps commanders, then they must keep in touch by despatching messengers. Communicating by telephone is cheating! All messages *must* be relayed through the umpire if the campaign is to be at all realistic: and umpires can do nasty things like getting your messenger captured or lost, thus spoiling the best laid plans as in real life!

The next question is that of supply. Of course, it is possible to assume at the beginning of the campaign that every unit is automatically in full supply all the time, but this is very unrealistic. In practice, the further troops move away from their supply depots the more vulnerable their lines of communication become, and a unit cut off from food supplies for more than a few days would suffer heavily, not just from malnutrition, but also from desertion.

Supplies can be introduced into a campaign quite easily, however. If each country has a specified number of supply bases, wagons can be dispatched automatically each day and assumed to automatically reach the troops they are supplying by nightfall. As the army moves further away from its supply base, of course, you will reach a situation where a long train of supply wagons, each a day apart, will be following each other down a well-trodden path . . . until an enemy unit interposes itself across their line of march and seizes all their supplies as they arrive!

This can be nasty, because it could be several 'days' before you are aware your supply line has been cut, by which time your troops have moved further away from the enemy blocking unit, and the first indication you have of the situation is when the umpire informs you that your wagon train failed to arrive one evening.

A certain amount of food and fodder could usually, of course, be obtained from local villages, so rule that when this happens units can forage for two days. At the end of that two days, if the supply lines have not been re-opened, five per cent of your force deserts. A further five per cent will desert each day for a further week unless supplies get through: by the end of this time your units are down to 65 per cent of their original strength, and starving hungry; their morale suffers a -5 and their daily rate of march is halved. During the following days the units lose ten per cent per day each day their supply lines remain closed, and suffer a further -1 from their morale each day. In six days the force has completely evaporated.

This may sound fairly drastic but it is a good way of making people aware of the supply problems, makes them guard their lines of communication (something which very rarely enters into the one-off battle situation), makes them employ proper cavalry scouting and screening procedures, and enjoins great caution.

The major exception to all the above, of course, is the French Army which, throughout the Napoleonic Wars, was forced to live off the land rather than rely on cumbersome wagon trains. Thus for the French we have to have different rules.

Each grid square on your campaign map (apart from supply depots which have unlimited capacity) is capable of supplying five battalions of infantry, 15 squadrons of cavalry or five batteries of guns per day. This is collected by figures–one from each unit–who are despatched foraging each day. They can, obviously, only forage from squares adjacent to their army's line of march or within reach during a single day.

This forces the French player to split his army up into a large number of fairly small units marching independently–as in historical practice they did–and forces the French player to plan his campaign so that all units can be kept in supply and yet at the same time be within fairly close call of each other so that they can concentrate to join battle.

And just as non-French supply lines can be cut, so the French player can be made to suffer for his apparent ease of movement: any enemy unit moving into a grid square where troops are foraging automatically kill them and capture the foraged supplies–thus the French player can also suddenly find himself without food! However, French soldiers were equipped with *four* days' iron rations, and can survive that long before the five per cent attrition effect comes into force.

The next question is that of recruitment and casualty replacement, a tricky one because if the recruitment rate is too high you can soon reach a state where everybody involved is madly painting figures every evening, and there still aren't enough to go round!

I have found the best way round this is merely to have *replacements,* plus bonus recruitments. Every time a battle is fought, sufficient recruits to replace the casualties are mustered in each player's capital: they then have to march the full distance to join up with their army. This means that if an army is forced to fight two engagements within a couple of days of each other, its strength is already reduced at the beginning of the second battle.

In addition to this, the *winning* player is allowed to choose *one* additional unit to join his army: this may be a line infantry battalion, line cavalry regiment, or line artillery battery, at the player's discretion.

Similarly, a victory adds one morale point to the basic morale factor of all units involved in that battle next time they fight, while a defeat subtracts one morale point.

Units which manage to survive through three battles, however, whether they win or lose, become veterans after the third action. Guard units can only have casualties made up from veteran units–which means each player needs to carefully conserve his guards until at least the third battle unless he wants his most effective fighting force disastrously weakened.

One factor which will inevitably enter into a wargames campaign is engineering work–building bridges and redoubts in particular. Special units of pontonniers and engineers can be attached to armies for this purpose if desired, or alternatively a battalion of infantry can be assigned to a task, a group of engineers being assumed to be in charge. A simple chart can be drawn up for these activities in any case, as follows:

Bridging a river takes half a day for every sixteenth of an inch width on the map–thus a river an eighth of an inch wide will take a full day to bridge, etc.

To destroy or mine any bridge takes half a day.

To dig a trench takes half a day per man–ie, a whole battalion digging themselves in will take half a day, and so on. Filling in a trench is easier and need only take a quarter of a day.

Building redoubts takes longer than digging trenches, so allow one day per battalion to construct a redoubt for one battery. A redoubt can be pulled down in half a day.

Making gabions, fascines etc takes the same time as digging a trench. (Note: only properly constructed redoubts should count as 'hard' cover for artillery. Gabions may be used to give guns protection but should only count as 'soft' cover).

Weather is another factor which can play an important part in a campaign, but since this has been dealt with very thoroughly in other publications (notably Donald Featherstone's *Advanced War Games*) I will simply say that allowance should be made for mud and rain hampering progress at certain times of the year, while heavy snow will obviously put a severe strain on the French foraging situation.

Objectives: there is no point in a campaign unless it has an ultimate aim. One possibility is giving the French player the task of capturing all his opponent's capital cities, but this is both unrealistic and practically impossible, since all the other players need to do is sit tight at home and wait for the French to come to them. Another is the destruction of all the opposing armies–difficult because casualties in a battle rarely exceed 30 per cent, and because of the replacement situation.

However, a similar concept can be designed which works well, whereby any nation which suffers two *consecutive* defeats retires from the campaign and becomes a neutral–except France, of course. This means that a French 'blitzkrieg' against one opponent can eliminate him in fairly short order, reducing the odds against the French player.

In passing, it is worth pointing out that the French Army should not be made too numerically superior to its opponents, or the battles become bloodbaths pure and simple with little tactical finesse. Ideally, the French Army should be made sufficiently strong to be able to defeat any one of its opponents in a straightforward battle, but conversely if any two of its opponents succeed in joining forces they should be strong enough to beat the French.

In order to stop the Allies simply linking up *all* their forces and marching to invade France, a couple of restraining factors need to be introduced. One is to say that, if a player's capital is captured, he is automatically eliminated from the campaign. This means that at least a proportion of each force will have to be held in reserve to protect the city. Another would be to say that no Allied player may send more than three-quarters his total force to the aid of another country, but must keep a quarter in his own territory. A third possibility is to give each Ally a -1 on the morale of all units fighting outside their own country. Other ideas will no doubt occur to you, but keep in mind that the main aim is the *enjoyment* of the game: too much freedom and the campaign will be over in a couple of weeks; too many restrictions and the players will become frustrated and bored.

Many other factors could be taken into account, ranging from sieges (which I have not discussed because they do not generally work very well as wargames) to technical invention: suppose Napoleon had had balloons at Waterloo, for example? Or breech-loading rifles? This factor could be introduced using a chance card system, whereby at the beginning of each campaign week the umpire draws a chance card: this could either be blank or could give some new technical or tactical innovation to one player or another. The technical innovations should be kept within bounds, however–it wouldn't be much of a Napoleonic campaign with Tiger tanks hiding in La Haye Sainte or Phantoms roaring overhead!

Nevertheless, I hope the ideas in this chapter may have sparked off your imagination sufficiently and given enough information for you to begin organising your own campaign.

A French Hussar and Chasseur a Cheval de la Garde 'discuss' the battlefield situation (or possibly the chances of leave!) during a battle (Donald Featherstone).

FRENCH **Troop type**	**Fire factor**	**Melée fac**	
		Impact	Co
Infantry			
Old Guard	3 : 3	4 : 5	4
Young Guard	- : 2	- : 3	-
Grenadier	2 : 2	3 : 3	3
Carabinier	3 : 3	3 : 3	3
Fusilier	1 : 1	2 : 2	1
Chasseur	1 : 1	2 : 2	1
Voltigeur	2 : 2	3 : 3	2
Dragoon	0 : 0	2 : 2	1
Cavalry			
Guard heavy	-	4 : 5	5
Cuirassier	-	3 : 4	4
Dragoon	0 : 0	2 : 3	3
Guard light	1 : 1	2 : 3	3
Hussar	0 : 0	1 : 2	2
Chasseur	0 : 0	1 : 2	2
Guard lancer	-	- : 5	-
Lancer	-	- : 4	-
Artillery			
Guard foot	- : 3	-	-
Guard horse	2 : 3	-	2
Foot	1 : 2	-	1
Horse	1 : 2	-	0

appendix one

Playing rules

1 model soldier represents 33 real men.
1 mm represents 1 yard in horizontal scale.
1 contour of hill, tree or house is 40 feet high.
1 game Move represents 2½ minutes of real time.

Orders

Each player should write out orders for each unit under his command at the beginning of the game. These should state objectives, initial positioning etc. Unit commanders have freedom of action within the broad scope of these orders. Drastic changes of plan must be sent to the unit in question by the commanding officer via messenger. Changes of orders take half a game move to write and half a game move to read. Messengers move at light cavalry charge speed all the time if required.

Units may not be given orders in violation of the tactics or knowledge available during the Napoleonic Wars or alien to their troop type (eg, line infantry cannot be ordered to act as skirmishers).

When the commander-in-chief figure is actually with a unit, changes of orders may be given verbally and can be acted upon instantly.

National characteristics

The tables following give all the basic

		Movement factors							
rale	Control	Line	Column	Road	Skirmish	Uphill	Downhill	Woods etc.	Charge
10	0 : 0	150	180	200	-	- 10	- 15	- 50	+ 30yds
9	- : 1	140	160	180	180	- 15	- 20	- 60	+ 25
9	1 : 1	140	160	180	-	- 15	- 20	- 60	+ 20
9	1 : 1	140	160	180	180	- 15	- 20	- 50	+ 25
7	3 : 3	120	140	160	-	- 25	- 30	- 75	+ 20
7	3 : 3	120	140	160	160	- 25	- 30	- 70	+ 20
8	2 : 2	130	150	170	170	- 20	- 25	- 60	+ 25
6	3 : 3	110	130	150	130	- 25	- 30	- 75	+ 15
10	0 : 0	280	320	360	-	- 40	- 60	- 100	+ 60
9	1 : 1	270	310	350	-	- 50	- 70	- 120	+ 50
8	1 : 1	280	320	360	320	- 50	- 70	- 120	+ 50
10	1 : 1	360	400	450	400	- 30	- 50	- 80	+ 80
8	4 : 3	350	390	440	390	- 40	- 60	- 100	+ 70
8	3 : 2	350	390	440	390	- 40	- 60	- 100	+ 70
10	- : 1	360	400	450	-	- 40	- 60	- 100	+ 80
8	- : 2	350	390	440	-	- 50	- 70	- 120	+ 70
		Cross-country		Road	Uphill	Downhill	Woods etc.		Charge
10	- : 0	180		200	- 20	- 30	Not allowed		-
10	1 : 1	310		350	- 40	- 60	- 100		+ 50
8	1 : 1	140		160	- 25	- 35	Not allowed		-
9	2 : 2	290		330	- 45	- 65	- 110		+ 40

information needed about the basic types of troops employed during the Napoleonic Wars. If using a unit which is not listed, take a comparable type and use those characteristics. An explanation of the various factors follows.

Fire factor This is a composite figure reflecting both the accuracy of the weapon being used and the capabilities of the unit in question, its state of training etc. Use this factor in conjunction with the rules for firing.

Impact factor This is the factor used by any unit *charging* into contact. It is used during the first round of a melée only. Use this factor in conjunction with the melée rules.

Confused melée factor This is the factor used in all subsequent rounds of melée.

The two factors are different in many cases, which represent the different abilities of troop types (eg, some are more effective when charging in formation, others are superior in a confused hand-to-hand fight). They also represent, again, the state of a troop type's training and equipment.

Morale factor This is a numerical evaluation given to each troop type's basic morale as a reflection of its status, training and equipment. Use in conjunction with the morale rules.

Control factor This factor represents a unit's responsiveness to control by its officers, and is related to its morale factor. Use in conjunction with the control rules.

Movement factors These represent in yards the furthest distance a unit may move depending on its formation and the type of terrain it is crossing. Note that only certain troop types may act as skirmishers, while most cavalry would not be expected to use their carbines at all. The reason the penalty for moving downhill is higher than that for moving uphill is because it is more difficult for infantry and cavalry to maintain a formation when moving downhill. Similarly, artillery is slower because the gun team have to prevent the guns running forward onto the horses. Use in conjunction with the movement rules.

PRUSSIAN	Fire	Melée facto	
Troop type	**factor**	Impact	Conf
Infantry			
Guard	2 : 3	3 : 3	3 :
Grenadier	1 : 2	2 : 2	2 :
Musketeer	0 : 1	1 : 2	1 :
Fusilier	1 : 1	1 : 2	1 :
Jäger	1 : 2	1 : 2	1 :
Landwehr	- : 0	- : 0	- :
Cavalry			
Cuirassier	-	4 : 3	5 :
Dragoon	0 : 0	3 : 2	4 :
Uhlan	-	4 : 3	3 :
Hussar	0 : 0	2 : 1	3 :
Horse jäger	0 : 0	2 : 1	3 :
Artillery			
Foot	1 : 2	-	1 :
Horse	1 : 2	-	0 :

BRITISH	Fire	Melée facto	
Troop type	**factor**	Impact	Conf
Infantry			
Foot Guard	4 : 4	4 : 4	5 :
Fusilier	3 : 3	3 : 3	4 :
Grenadier	3 : 3	2 : 2	3 :
Line	2 : 2	1 : 1	2 :
Light	3 : 3	3 : 3	2 :
Highlander	2 : 2	3 : 3	4 :
Rifle	0 : 0	1 : 1	2 :
Cavalry			
Life & RH Guard	-	5 : 5	6 :
Dragoon Guard	1 : 1	4 : 4	5 :
Dragoon	0 : 0	3 : 3	4 :
Light dragoon	0 : 0	2 : 2	3 :
Hussar	0 : 0	2 : 2	3 :
Artillery			
Foot	2 : 2	-	2 :
Horse	2 : 2	-	1 :

le	Control	Movement factors							
		Line	Column	Road	Skirmish	Uphill	Downhill	Woods etc.	Charge
9	0 : 0	110	130	150	-	- 20	- 30	- 60	+ 20yds
8	1 : 0	100	120	140	-	- 20	- 30	- 60	+ 20
6	2 : 2	80	100	120	-	- 25	- 35	- 70	+ 15
7	2 : 2	80	100	120	120	- 25	- 35	- 60	+ 15
8	2 : 2	100	120	140	140	- 20	- 30	- 50	+ 20
4	- : 4	60	80	100	-	- 30	- 40	- 60	+ 15
7	0 : 1	270	310	350	-	- 50	- 70	- 120	+ 50
6	0 : 1	280	320	360	320	- 50	- 70	- 120	+ 50
6	2 : 3	350	390	440	-	- 50	- 70	- 120	+ 70
6	3 : 4	350	390	440	390	- 50	- 70	- 120	+ 70
7	3 : 4	360	400	450	400	- 50	- 70	- 120	+ 70
		Cross-country		Road	Uphill	Downhill	Woods etc.		Charge
7	0 : 1	100		120	- 30	- 40	Not allowed		-
8	1 : 2	180		220	- 50	- 70	- 110		+ 30

Note On all factors two numbers are given separated by a colon (eg, 9:8). This represents the different capabilities of different troop types before 1808 and after 1808. 1808 has been chosen as a convenient 'watershed' because most

ale	Control	Movement factors							
		Line	Column	Road	Skirmish	Uphill	Downhill	Woods etc.	Charge
10	0 : 0	140	170	185	-	- 15	- 20	- 60	+ 25yds
9	1 : 1	130	160	175	160	- 15	- 20	- 50	+ 25
8	1 : 1	120	150	165	-	- 20	- 25	- 70	+ 20
7	2 : 2	110	140	155	-	- 25	- 30	- 80	+ 15
8	2 : 1	120	150	165	165	- 20	- 25	- 60	+ 20
9	3 : 3	110	140	155	-	-25	- 30	- 70	+ 20
9	2 : 2	130	160	175	160	- 15	- 20	- 50	+ 25
10	1 : 1	270	310	350	-	- 40	- 60	- 100	+ 60
9	2 : 2	280	320	360	320	- 40	- 60	- 100	+ 60
8	4 : 4	270	310	350	310	- 50	- 70	- 110	+ 50
7	5 : 5	340	380	430	380	- 40	- 60	- 100	+ 70
7	6 : 6	340	380	430	380	- 40	- 60	- 100	+ 70
		Cross-country		Road	Uphill	Downhill	Woods etc.		Charge
7	2 : 2	140		155	- 25	- 35	Not allowed		-
8	3 : 3	250		310	- 45	- 65	- 100		+ 40

reorganisation took place in European armies between 1807-8. When refighting a historical battle, take care to use the correct factors. When fighting a fictitious battle either factor may be used at the player's discretion. In cases where a troop type did not exist before 1807-8, I have left a dash (-). So you can't have French lancers at Austerlitz!

Weather

Throw one average dice at the start of the game. 2= heavy rain; 3= changeable; 4 and 5 =fine. In heavy rain all movement and fire effectiveness is halved (except that troops in buildings may fire as normal). Heavy rain also puts out all fires after two game moves. If a 3 is thrown (changeable), dice for weather again at end of third game move: 2= heavy rain; 3 =light rain; 4=still changeable; 5= fine. In light rain all movement and firepower is reduced by a quarter. Fires are put out after four game moves. If a 4 has been thrown again, dice for weather a third time at end of sixth game move. Players may wish to dice for weather changes other than the above (eg, to find out whether heavy rain present at the beginning disperses later). This is entirely up to you. When refighting a historical battle, of course, the weather applicable to the actual day should be used.

Wind

Wind strength and direction are important in deciding how fast smoke moves and disperses, and how far and fast fires spread. Throw another average dice at the beginning of the game. 2= no wind, smoke lasts 5 moves; 3= light wind, smoke lasts 3 moves; 4 = medium wind, smoke lasts 2 moves; 5 =strong wind, smoke lasts 1 move. Strong wind blows smoke 300 yards per move; medium wind 200 yards; light wind 100 yards. Fire spreads 25 yards per move through any inflammable material (wooden houses etc) in direction wind is blowing. Wind direction may be decided at start of game in any way the players desire.

Visibility

All units on a wargames table, except those hidden by woods, hills or buildings, are considered to be visible the full length of the table except under adverse

RUSSIAN Troop type	Fire factor	Melée fact	
		Impact	Conf
Infantry			
Guard grenadier	2 : 2	4 : 4	5 :
Guard jäger	2 : 2	3 : 3	3 :
Grenadier	1 : 1	3 : 3	4 :
Musketeer	0 : 0	2 : 2	3 :
Jäger	1 : 1	2 : 2	3 :
Cavalry			
Guard cuirassier	-	5 : 5	6 :
Guard dragoon	1 : 1	4 : 4	5 :
Cuirassier	-	4 : 4	5 :
Dragoon	0 : 0	3 : 3	4 :
Guard uhlan	-	5 : 5	4 :
Uhlan	-	4 : 4	3 :
Guard light	1 : 1	3 : 3	3 :
Hussar	0 : 0	2 : 2	2 :
Guard cossack	-	4 : 4	2 :
Cossack	-	3 : 3	1 :
Artillery			
Position	2 : 2	-	3 :
Light	2 : 3	-	2 :
Horse	3 : 3	-	1 :

AUSTRIAN Troop type	Fire factor	Melée facto	
		Impact	Conf
Infantry			
Grenadier	2 : 2	3 : 3	4 :
Fusilier	1 : 1	2 : 2	1 :
Light	1 : 1	1 : 2	0 :
Jäger	2 : 2	2 : 3	1 :
Cavalry			
Cuirassier	-	4 : 3	5 :
Dragoon	1 : 0	3 : 2	4 :
Uhlan	-	4 : 3	3 :
Hussar	1 : 0	2 : 1	3 :
Artillery			
Heavy	0 : 1	-	1 :
Light	0 : 1	-	0 :

e	Control	Line	Column	Road	Skirmish	Uphill	Downhill	Woods etc.	Charge
		Movement factors							
10	0 : 0	130	160	170	-	- 15	- 20	- 60	+ 25yds
10	1 : 1	140	150	170	170	- 15	- 20	- 50	+ 25
9	1 : 1	120	140	160	-	- 20	- 30	- 70	+ 20
8	2 : 2	100	130	150	-	- 25	- 35	- 80	+ 20
9	3 : 3	120	140	160	160	- 20	- 30	- 60	+ 25
10	0 : 0	260	300	340	-	- 50	- 70	- 130	+ 50
10	1 : 1	280	320	360	320	- 40	- 60	- 120	+ 60
9	1 : 1	250	290	330	-	- 60	- 80	- 140	+ 50
8	2 : 2	270	310	350	310	- 70	- 90	- 150	+ 60
10	2 : 2	350	390	440	-	- 40	- 60	- 120	+ 80
8	3 : 3	340	380	430	-	- 50	- 70	- 130	+ 70
10	2 : 2	350	390	440	390	- 40	- 60	- 120	+ 80
8	4 : 4	340	380	430	380	- 50	- 70	- 130	+ 70
9	3 : 3	340	380	430	400	- 30	- 50	- 100	+ 80
7	5 : 5	320	360	420	390	- 40	- 60	- 110	+ 70
		Cross-country		Road	Uphill	Downhill	Woods etc.		Charge
8	0 : 0	80		110	- 40	- 50	Not allowed		-
8	1 : 1	130		150	- 30	- 40	Not allowed		-
9	2 : 2	240		300	- 40	- 60	- 120		+ 40

weather conditions. This does not mean they are all *recognisable*, and since the presence of unidentified units can have a strong influence in a battle, it is important

e	Control	Line	Column	Road	Skirmish	Uphill	Downhill	Woods etc.	Charge
		Mpvement factors							
7	1 : 1	120	140	160	-	- 15	- 20	- 60	+ 25yds
4	2 : 3	90	110	130	-	- 25	- 30	- 75	+ 20
5	3 : 3	90	110	130	130	- 25	- 30	- 70	+ 20
7	2 : 2	120	140	160	160	- 25	- 30	- 60	+ 25
7	0 : 1	270	310	350	-	- 50	- 70	- 120	+ 50
6	0 : 1	280	320	360	320	- 50	- 70	- 120	+ 50
6	2 : 3	350	390	440	-	- 50	- 70	- 120	+ 70
6	3 : 4	350	390	440	390	- 40	- 60	- 100	+ 70
		Cross-country		Road	Uphill	Downhill	Woods etc.		Charge
6	2 : 1	110		130	- 30	- 40	Not allowed		-
7	3 : 2	180		220	- 50	- 70	- 110		+ 30

Observer is on:	Level ground	1 contour	2 contours	3 contours	4 contours
Obstruction:	**Visibility is obscured for:**				
1 contour	X	X	200 yards	150 yards	100 yards
2 contours	X	X	X	150 yards	100 yards
3 contours	X	X	X	X	100 yards
4 contours	X	X	X	X	X
Skirmishers	50 yards	25 yards	0	0	0
Infantry unit	100 yards	50 yards	25 yards	0	0
Cavalry unit	150 yards	75 yards	30 yards	10 yards	0
Smoke from burning buildings etc.	X	150 yards	75 yards	30 yards	10 yards

(Remember a tree or house is the equivalent of one contour high)
Note: 'X' = no visibility beyond the obstruction at all.

to know at what distances they are recognisable. Units outside recognisable distance may be represented on the table by markers instead of figures if the players so wish.

The distances at which units are recognisable are as follows: single figure–50 yards; infantry company or cavalry squadron–200 yards; infantry battalion–800 yards; cavalry regiment–1000 yards; larger units–1500 yards. Officers with telescopes can recognise units at twice the above distances. Heavy rain reduces overall and recognisable visibility by ½, light rain by ¼.

Smoke and mist

Smoke collects in front of an infantry unit or artillery battery after three consecutive rounds of firing, and reduces visibility to half the above values while it lasts. Targets which are no longer visible due to smoke obscuring the aim may not be fired upon again until the smoke clears. The height of a cloud of smoke is taken as the equivalent of half a contour, so troops or guns on a hill can fire over smoke at a lower level.

Early morning mist was a feature of many Napoleonic battlefields (eg, Austerlitz in particular) and players may wish to recreate this on the tabletop. The simplest way of doing this is to use markers for all troops hidden in mist (ie, in valley bottoms or alongside rivers) and reduce visibility by three-quarters. Mist gradually disperses during three moves, visibility being reduced by half in the second and a quarter in the third. In medium wind mist disperses after two moves; in strong wind after one move.

Dead ground

Areas of dead ground are formed behind all obstacles on the wargames table, and targets in dead ground may not be fired upon. Remember, all firing in this period was by line of sight only: there was no indirect fire.

Movement

Apart from basic terrain, movement can be affected by various other factors, which must all be taken into account. These are as follows:
To cross a hedge, fence, ditch or low wall, deduct 1/4 move for infantry and cavalry. Artillery can only pass through gates, gaps and bridges.
To cross a stream deduct 1/3 move from all units.
To cross a fordable river deduct 1/2 move. All units fording rivers become unformed and must reform on the other side. Artillery can only cross by bridges.
To enter or leave a building deduct 1/2 move. All troops except light infantry are unformed in buildings and must reform on leaving. Mounted cavalry and artillery may not enter buildings.
No unit may charge when moving downhill, across rivers (except by bridges), or into buildings. The last 50 yards of infantry move or 150 yards of cavalry move must be in a straight line when charging. *This is also the minimum distance needed to reach charge speed,* (ie, troops 'charging' over

shorter distances may not use their impact factor).

Tired troops deduct 1/2 move from their movement factor. Exhausted troops may not move at all. Tired troops moving become exhausted.

To turn an artillery battery through 0-30° takes 1/5 move; 30-60° takes 3/5 move; 60-120° takes 1 move; 120-150° takes 1/2 move; 150-180° takes 1/4 move. Use a protractor when refacing batteries.

Changes of formation

Each take a specified portion of a move depending on nationality and type as follows:-

	Infantry			Cavalry	
	British French	Austrian Prussian	Russian	French Russian	Other
Line to column	1	1½	1½	½	1
Column to line	1	2	1½	½	1
Line to square	½	1	½	-	-
Square to line	½	1½	1	-	-
Column to square	¼	¾	½	-	-
Square to column	¼	¾	½	-	-
Line to skirmish	½	1	¾	½	¾
Skirmish to line	¾	1½	1	1	1½
Column to skirmish	¾	1½	1	¾	1
Skirmish to column	1	2	1½	1	1½
Reform when unformed	½	1	¾	1	1½
Reform when disordered	1	2	1½	2	3
Unlimber a battery	⅓ Foot	1 Foot	½ Foot	⅕ Horse	½
Limber a battery	½ Foot	1½ Foot	¾ Foot	⅓ Horse	¾

Conscripts and militia (including Landwehr) take double above times.
All troops are unformed while changing formation.

Reaction to enemy movement

A unit may not react or take any action to counter movements of an enemy unit which it cannot see. When enemy movement becomes visible, a quarter move reaction time must elapse before any action can be taken, *except that infantry may begin to form square the moment enemy cavalry movement is visible.* Infantry menaced by cavalry *must* attempt to form square unless both flanks are already secure and cannot be outflanked.

Troops attacked from the flank *must* attempt to turn and face the threat. A line must try to wheel to meet the enemy, but troops in column can simply face through 90° (takes 1/10 move). Troops charged before their change of front is complete fight unformed.

Cavalry must always try to counter-charge enemy cavalry. They should attempt to wheel to face the direction of the enemy charge, but if this would leave them insufficient time to get up charge speed (ie,150 yards), they may charge unformed. They are then disordered at the end of the first round of melée.

Skirmishers may not approach closer than 40 yards to *formed* enemy troops and must retire at least this distance before a formed advance. If charged in the open skirmishers may fire one volley (1/5 effect) and then run towards nearest cover or friendly troops. If caught they will have their backs to the enemy and are classed as routed.

Control

A control test is taken under the following circumstances: after winning a melée, or when the unit commander is killed, or when the unit commander moves more than 300 yards away from the unit, or after a charge in which a unit fails to make contact.

Throw one average dice and add the unit control factor as given in the national characteristics charts, then add or deduct as follows:

Enemy retiring within 300 yards +1; retreating +2; routing +3; standing -1; advancing -2; advancing to flank or rear -3.

Friends retiring within 300 yards -1; retreating -2; routing -3; standing +1; advancing +2.

For each round of melée won by unit under test within four moves +1.

Unit commander within 150 yards + or -1; commander-in-chief within 150 yards + or - leadership factor.

Unit is veteran + or -1.

Unit is under fire from flank or rear -1.

Note that all these factors are cumulative. '+ or -' factors may be chosen to give the best result possible.

If total comes to 13 or over, unit will pursue or charge towards nearest enemy unit(s) for two moves; 11-12, pursue or charge towards nearest enemy for one move; 5-10 reform and then act as own commander wishes; 2-4 reform and then may not *advance* for one move; under two, retire towards nearest friends for one move and then reform.

Small arms and artillery fire

1. All fire is by line of sight only. In other words, the unit or battery firing must be able to trace an unobstructed straight line between themselves and the target.

2. Only howitzers may fire over the heads of intervening troops, and then only at targets which are at least one contour higher than both the gun firing and the intervening men. *Exception:* guns on hills may fire ball (and howitzers shell) over the heads of intervening troops on a lower level. Similarly, infantry on a higher contour may fire over the heads of troops on a lower level, subject to the dead ground rules. Canister may *not* be fired over the heads of intervening troops at any time.

3. Weapons may be fired at woods or buildings which are believed to house enemy troops. However, the opposing player is under no obligation to say whether there are actually any troops present. Similarly, if the opposing player is able to withdraw the hidden troops thus under fire without bringing them into his opponent's view, he may mark such a move on his order sheet and not inform his opponent that they have been moved until such time as they re-appear in sight. Even if the firing player suspects something like this is going on, he must continue to fire at the houses or wood either (a) until the deception is revealed or (b) the troops firing are threatened by a more immediate target.

4. In cases where a unit is only able to fire for part of a move, or when a target is only visible for part of a move, refer to the chart on page 34.

Artillery ranges	**Canister**			
Type of battery	**+4**	**+2**	**0**	**-2**
3/4pdr	0– 75	- 150	- 250	- 3
6/7pdr	0– 85	- 165	- 275	- 4
8pdr	0–100	- 200	- 325	- 4
9pdr	0–120	- 225	- 350	- 4
12pdr	0–150	- 300	- 400	- 5
5.5" howitzer	0– 50	- 100	- 150	- 2
6/7" howitzer	0– 70	- 125	- 200	- 2
8" howitzer	0–100	- 225	- 375	- 5

Infantry weapon ranges	0 - 35
Musket	+3
Mounted carbine	0
Dismounted carbine	+1
Rifle	0

Small arms and artillery fire—add or de

Firing unit is	Musket Carbine	Rifle	3/4 B
Resting weapons	+1	+1	
Unformed	- 2	- 2	- 2
Disordered	- 4	- 4	- 4
Target is			
Behind soft cover	- 3	- 2	- 2
Behind hard cover	- 4	- 3	- 3
In skirmish order	- 4	- 2	- 4
Lying down	- 3	- 2	- 3
+1 for every –	3	3	4

Unit firing is veteran + 1. Conscript - 2. First

5. Only the front rank of a unit may fire unless, of course, the rear ranks are on higher contours.

Note

The whole of a target must be behind cover to qualify for savings.

The additions for number of ranks in the target unit only extends as far as the firing weapon's penetration ability, see range chart.

Player firing must nominate type of ammunition being used *before* measuring range.

Penetration is halved after one move of heavy rain, two moves of light rain, or against targets on the facing slope of a hill, in woods, outbuildings or marshy ground, except for infantry weapons.

Skirmisher screens When firing at a unit protected by a skirmisher screen, deduct for skirmish order but add on for ranks deep. This reduces the casualties and simulates the harassing effect of skirmisher fire on a defender's accuracy.

Firing

Throw one fire dice (see Chapter one) and add the fire factor for the unit firing. Then add or deduct according to the range and circumstances as detailed in the following charts:

	Ball or Shell	Will penetrate target to a depth of:
	+2 0 -3 -6 -9	
Range in yards	0—100 - 275 - 530 - 770 - 900	150 yards
	0—125 - 300 - 580 - 820 - 950	175
	0—150 - 350 - 660 - 880 - 1000	200
	0—175 - 380 - 720 - 910 - 1050	210
	0—225 - 450 - 800 - 1000 - 1250	250
	0—200 - 400 - 750 - 950 - 1100	50
	0—225 - 425 - 775 - 975 - 1150	70
	0—250 - 450 - 800 - 1000 - 1250	100
	- 125 - 150 - 175 - 200 - 300 yards	
	2 -4 -6 -9 -	50
	6 -9 - - -	50
	4 -6 -9 - -	50
	3 -4 -5 -6 -9	50

propriate and compare final score with tables overleaf

dr :	8 pdr B C	9 pdr B C	12pdr B C	5.5" S C	6/7" S C	8" S C	B= Ball C= Canister S= Shell
2	-3 -3	-3 -3	-3 -3	-3 -3	-3 -3	-3 -3	
4	-5 -5	-5 -5	-5 -5	-5 -5	-5 -5	-5 -5	
							Note: British guns only may fire shrapnel instead of shell, and count an extra +1 against targets in the open.
3	-1 -2	-1 -2	0 -1	0 -2	0 -2	0 -1	
4	-2 -3	-2 -3	-1 -2	0 -3	0 -3	0 -2	
3	-5 -4	-5 -4	-6 -5	-4 -3	-4 -3	-3 -2	
1	-4 -1	-4 -1	-5 -2	-2 -1	-2 -1	-1 0	
5	3 4	3 4	2 3	2 4	2 3	1 3	ranks deep of target

ing + 2. Small arms v. cuirassiers - 1. **NB** only troops in hard cover can rest weapons.

Score	Number of figures firing (small arms): or : Melée head count															
	1	2	3	4	5	6	7	8	9	10	11	12	13	14	15	16
- 10	1	1	1	1	1	1	1	1	1	1	2	2	2	2	2	2
- 9	1	1	1	1	1	2	2	2	2	2	3	3	3	3	3	4
- 8	1	1	1	1	1	2	2	2	2	3	3	3	3	4	4	4
- 7	1	1	1	2	2	2	3	3	3	4	4	4	5	5	5	6
- 6	1	1	2	3	3	4	5	5	6	7	7	8	9	9	10	11
- 5	1	2	3	4	5	6	7	8	9	10	11	12	13	14	15	16
- 4	1	3	4	5	7	8	9	11	12	13	15	16	17	19	20	21
- 3	2	3	5	7	8	10	12	13	15	17	18	20	22	23	25	27
- 2	2	4	6	8	10	12	14	16	18	20	22	24	26	28	30	32
- 1	2	5	7	9	12	14	16	19	21	23	26	28	30	33	35	38
0	3	5	8	11	13	16	19	21	24	27	29	32	35	37	40	43
+ 1	3	7	10	13	17	20	23	27	30	33	37	40	43	47	50	53
+ 2	4	8	12	16	20	24	28	32	36	40	44	48	52	56	60	64
+ 3	5	11	16	21	27	32	37	43	48	53	59	64	70	75	80	85
+ 4	7	13	20	27	33	40	47	53	60	67	73	80	87	93	100	107
+ 5	8	17	25	33	42	50	58	67	75	83	92	100	108	117	125	133
+ 6	10	20	30	40	50	60	70	80	90	100	110	120	130	140	150	160
+ 7	12	23	35	47	58	70	82	93	105	117	128	140	152	163	175	187
+ 8	13	27	40	53	67	80	93	107	120	133	147	160	173	187	200	213
+ 9	17	33	50	67	83	100	117	133	150	167	183	200	217	233	250	267
+10	20	40	60	80	100	120	140	160	180	200	220	240	260	280	300	320

For numbers in excess of 16 men firing, add two scores to the right total (eg, 24 figures firing, take the score for 12 and double it). **Note:** *all casualties men, not figures.*

Casualties from artillery fire

No of guns in battery	Score																				
	-10	-9	-8	-7	-6	-5	-4	-3	-2	-1	0	+1	+2	+3	+4	+5	+6	+7	+8	+9	+10
8	2	4	4	6	11	16	21	27	32	38	43	53	64	85	107	133	160	187	213	267	320
7	2	3	4	5	9	14	19	23	28	33	37	41	56	75	93	117	140	163	187	233	280
6	2	3	3	4	8	12	16	20	24	28	32	40	48	64	80	100	120	140	160	200	240
5	1	2	3	4	7	10	13	17	20	23	27	33	40	53	67	83	100	117	133	167	200
4	1	2	2	3	5	8	11	13	16	19	21	27	32	43	53	67	80	93	107	133	160
3	1	2	2	2	4	6	8	10	12	14	16	20	24	32	40	50	60	70	80	100	120
2	0	1	1	2	3	4	5	7	8	9	11	13	16	21	27	33	40	47	53	67	80
1	0	0	1	1	1	2	3	3	4	5	5	7	8	11	13	17	20	23	27	33	40

Special artillery rules

1. An artillery battery deployed for firing counts as a unit in skirmish order when fired upon.

2. A foot artillery battery when limbered counts as a normal close-order target.

3. A horse artillery battery when limbered counts as a unit in skirmish order.

4. A deployed battery fired on from right angles down its flank counts as a close-order target +2.

5. An artillery battery's effectiveness is reduced by one gun for every 10% casualties suffered.

6. Casualties on *limbered* batteries are taken equally from gunners and draft horses. When one horse figure (ie, 33 horses) is killed, the battery must halt *unformed* for one move to cut dead horses free and re-harness limbers.

Morale

A morale test must be taken under the following circumstances: infantry charged by cavalry (remember, count as unformed if in the process of forming square); infantry charged by infantry; infantry charging infantry; cavalry coming under fire during a charge; after each round of losing a melée; unit with friends retreating or routing within 300 yards *and in sight*; unit coming under fire for the first time (*not* veterans); unit suffering 10% casualties or greater in one Move; commander in chief or unit commander killed within sight of unit; when unit first falls below 50% strength.

Definitions

Unformed Unit in a melée or changing formation. Unit entering or leaving houses, crossing a river, hedge, ditch or wall. Line troops in buildings. Retreating units. Units pursuing routed troops. Units making an uncontrolled charge (following a control test only).

Disordered Unit after two rounds of melée, or first round of melée if unformed at start. Line troops in woods. Routing troops.

Formed Unit maintaining a recognised formation (includes light infantry in skirmish order).

Neither formed nor unformed Light infantry in houses or woods. Cossacks after one round of melée (Cossacks become unformed after two rounds, disordered after three).

Supported To be supported a unit must have at least one *formed* friendly unit of the same type facing the same front within 300 yards (Note: only infantry can support infantry and only cavalry can support cavalry, but artillery may be supported by either).

Unsupported No friendly troops within 300 yards.

Neither supported nor unsupported Unit with *unformed* friendly troops within 300 yards, or friendly troops within 300 yards but not facing the same front.

Secure flank Unit with flank against impassable object or with flank within 25 yards of another formed friendly unit. Infantry in square automatically have two secure flanks. Deployed artillery batteries with flanks resting on infantry units (especially if in square) also have secure flanks.

Threatened flank or rear Unit with insecure flank or rear within a charge move by any enemy unit, regardless of whether enemy is actually charging or not.

Flank neither secure nor insecure Unit which does not fulfil the conditions for a secure flank but is not threatened either.

Victory Unit which has won a round of melée, caused greater casualties in a fire fight, or broken an enemy's morale by a charge.

Defeat Unit which has lost a round of melée, suffered greater casualties in a fire fight, or had morale broken by an enemy charge.

Tired All units after one charge or rout move.

Exhausted All units attempting to move while tired become exhausted.

Tired troops must remain stationary and out of enemy fire for one move to return to normal. Exhausted troops must remain stationary and out of enemy fire for three moves to return to normal.

To take a morale test

Throw two average dice and add the difference between them to the unit's basic morale factor as laid down in the national characteristic charts. Add the leadership factor for any general officer within 50 yards of the unit and in sight. Then add or deduct for circumstances according to the following chart.

Morale chart

		Attack	Defense
Strength at start of move:	full	+2	+2
	7/8 to full	+1	+1
	¾ to 7/8	-1	-2
	5/8 to ¾	-3	-5
	½ to 5/8	-5	-7
	3/8 to ½	-7	-9
	¼ to 3/8	-9	-11
	Under ¼	-11	-15
Casualties suffered this move:	1/10	-1	-2
	1/8	-3	-5
	¼	-6	-8
	⅓	-9	-11
	½	-10	-15
	more than ½	-20	-30

(Continued on next page)

Morale – continued

		Attack	Defense
Formation:	formed	+2	+2
	neither formed nor unformed	0	0
	unformed	- 1	- 2
	disordered	- 3	- 5
Unit is:	supported	+1	+2
	neither supported nor unsupported	0	+1
Veterans: all guard and élite units (see also Chapter five). **Conscripts:** all militia and unseasoned troops.	unsupported	- 1	- 3
	out of sight of friends	- 3	- 4
	charging	+1	-
	conscript	- 1	- 1
	veteran	+1	+1
Does not apply to veterans	under fire for first time	- 3	- 2
Does not apply to veterans	under artillery fire	- 2	- 2
Does not apply to veterans	under fire from flank or rear	- 4	- 4
Does not apply to veterans	under fire from skirmishers	- 3	- 3
	tired	- 2	- 1
	exhausted	- 5	- 3
Unit has: Deduct if disadvantage	soft cover or higher ground	+1	+2
Deduct if disadvantage	hard cover	-	+4
	one secure flank	+1	+2
NB A square has two—not four—secure flanks	two secure flanks	+3	+5
	flanks neither secure nor insecure	0	+1
	threatened flanks	- 1	- 2
	for every victory scored within past four moves	+1	+2
	for every defeat suffered within past four moves	- 1	- 2
Unit receiving charge:			
	for every 3 ranks of attacking unit	- 1	- 2
Does not apply to British or veterans	first opened fire at range of 200 yards	-	- 1
Does not apply to British or veterans	150 yards	-	- 2
Does not apply to British or veterans	90 yards	-	- 3
Does not apply to British or veterans	60 yards	-	- 4
British and veterans - 2 only	35 yards	-	- 5
Friends within 300 yards forced back by enemy unit this Move		- 1	- 3
Unit on flank retreats or routs this Move		- 3	- 5
Unit's previous morale was good	(10 and over)	+1	+1
Unit's previous morale was poor	(6 or less)	- 1	- 1

Compare the final score with the following chart and act accordingly.

	Receiving a charge	Charging or countering	Advancing	Standing	Retiring	Retreating	Routing
10+	Advance	Charge	Advance	Advance	Halt	Retire	Retreat
7-9	Shaken	Advance	Halt	Halt	Retire	Retreat	Rout
4-6	Retire	Halt	Halt	Retire	Retreat	Rout	Rout
2-3	Retreat	Retire	Retire	Retreat	Rout	Rout	Rout
0-1	Rout	Retreat	Retreat	Rout	Rout	Rout	Rout
Below 0 - all rout							

Definitions

Charge Unit may complete charge move. Cavalry may only complete a charge against infantry in square if the infantry morale is 'shaken' or worse. Otherwise the cavalry flow round the square and become unformed.

Advance Unit may continue advancing, may stand or may retire at will. May also charge if required.

Halt Units advancing or retiring must remain stationary for one move. They may reform if unformed and carry on as normal in the following move.

Shaken Shaken units become unformed. Infantry or gunners firing deduct an additional -2 next Move.

Retire Infantry and foot artillery units retire 50 yards, remaining formed if they were already formed. Face towards enemy if attacked. Cavalry and horse artillery retire 100 yards. Gunners take guns with them unless involved in a melée.

Retreat Infantry and foot artillery retreat 100 yards, cavalry and horse artillery 200 yards. Backs to enemy if attacked. Gunners may attempt to limber their guns and take them along unless involved in a melée, when they will abandon them. Unformed.

Rout Infantry and foot artillery rout 250 yards, cavalry and horse artillery 500 yards. All units have backs to enemy if attacked. All units become disordered immediately. Gunners automatically abandon guns.

Line of retreat This is always towards nearest friendly troops, or cover, or the edge of the table, whichever is closest. Retreating units may pass through skirmisher lines but will flow round bodies of formed troops, becoming disordered in the process. Units facing a retreating unit must throw for morale themselves.

Rallying *Retiring troops* automatically rally at the end of one move of retirement, and reform if unformed. *Retreating troops* throw for morale again at the end of their first move of retreat unless under small arms fire, in which case they carry on retreating at normal move speed until out of range, then test their morale again. *Routing troops* throw for morale again when they have completed their first rout move, unless an enemy unit is within 400 yards, in which case they carry on routing at *charge* speed until the threat is removed. Units which rout off the table cannot be brought back into play, and any generals sent to help rally them are also removed permanently. Retreating and retiring units which move off the table may return when rallied.

Melees

1. A melée is formed when two bodies of troops move into physical contact.
2. Only the first two ranks of infantry and front rank of cavalry count during the first round of a melée; all troops involved count in the second and subsequent rounds.
3. Units with a wider frontage may overlap the opposing unit by one figure only on each flank during the first round of the melée.
4. Charging units use their impact factor *plus* their confused melée factor in the first round of a melée. All units use confused melée factors only in subsequent rounds.
5. No unit may *charge* into an already existing melée.
6. Head count. Each infantryman or gunner counts as one; each light cavalryman as two; each heavy cavalryman as three (charging lancers count as heavy cavalry in first round of melée). In other words, if you have 12 cuirassiers fighting 24 infantrymen, the cuirassiers count as 36 figures, the infantry as 24; similarly, 12 light cavalry versus 24 infantry would count as 24 to 24; 12 heavy cavalry versus 12 light cavalry as 36 to 24, and so on.
7. Troops in buildings or behind hard cover add 50% to their head count until such time as the enemy troops break through.
8. An infantry column or cavalry unit which wins the first round of melée against an infantry line breaks through the line and attacks the defenders in the rear in the second round of melée.
9. An infantry line which holds firm against an infantry column advances to flow round the column's flanks in the second round of melée.
10. A cavalry unit which breaks into an infantry square attacks the infantry in the back in the second round of melée.
11. Troops with backs to enemy must suffer casualties without fighting back.

Procedure Throw one fire dice and *deduct* from melée factor(s) of unit. Add or deduct according to following table. Read off final score against number of figures (ie, head count) on same table as for musket fire.

Additions and Deductions

Cavalry versus infantry in square, first round	-5
Infantry in square versus cavalry, first round	+2
Lancers versus infantry	+1
Cavalry versus infantry in one rank only	+2
Cavalry versus skirmishers	+6
Cavalry versus unformed infantry	+3
Cavalry versus disordered infantry	+4
Cavalry versus unformed cavalry	+1
Cavalry versus disordered cavalry	+2
Cossacks versus formed cavalry	-2
Infantry versus unformed troops	+1
Infantry versus disordered troops	+2
All troops following up	+1
All troops uphill from enemy	+1
Cavalry reinforced by a fresh squadron	+1
Tired troops	-2
Exhausted troops	-4
Unformed troops	-1
Disordered troops	-2

The winner of a melée is the side causing the greater number of casualties. Losers retire and throw for morale. If they stand winners must also throw for morale. If they retreat or rout winners need not take a morale test but must take a control test. If both sides stand, a second round of melée takes place. If winners get halt or worse in their morale test, losers must reform for one move before pursuing.

Buildings

A single model house, representing a group of buildings, can hold one infantry figure per centimetre of perimeter. In other words, if a house is 10 cm long by 6 cm wide, it can hold 10x2 + 6x2=32 figures. Figures need not be positioned inside the building but can be removed from the table and a simple note kept of their presence.

Stone buildings count as hard cover, wooden and half-timbered buildings as soft cover. Only wooden and thatched buildings may be set on fire. An infantryman can set a house on fire by remaining beside it for one move. Alternatively, buildings may be set on fire by shell fire (from howitzers only). After working out any casualties on troops in the house as normal, the firing player throws one average dice, and a score of 5 sets the house on fire.

Fires may be put out by heavy rain as outlined earlier, or by fire-fighting teams. These must be at least company strength. To put out a fire the fire-fighting troops throw one average dice, and a 4 or 5 puts it out. Troops engaged in fire-fighting may not engage in any other form of activity. Troops in a burning building who do not succeed in putting the fire out must evacuate after one move or are destroyed.

Protective values Each stone house or bridge has a protective value equal to 1 point per centimetre of wall. These points can be destroyed by artillery fire (ball or shell) as follows: after working out casualties on any troops in the building, the player firing throws a fire dice and to this adds the gun's fire factor. The total represents the number of protective points destroyed. The number of figures the house can hold is reduced by the same number (for example, in the 32-figure building referred to above, if five protective points were destroyed, five figures would have to leave). When half the building's protective points are destroyed it is considered to be no more than a heap of rubble, and further artillery fire is not effective.

Wooden buildings have a protective points value of 10; wooden bridges of 5.

appendix two

Miscellaneous

Model soldier manufacturers

Airfix Products Limited, Haldane Place, Garratt Lane, London SW18 4NB. 20 mm polythene figures of: French Old Guard and line infantry, cuirassiers and artillery; British line infantry and highlanders, hussars and horse artillery. Many other figures suitable for conversion to Napoleonic types. Cannot supply direct to the public, but figures available from Woolworths and most model shops.
Garrison Greenwood & Ball Ltd, 61 Westbury Street, Thornaby-on-Tees, Teeside. 25 mm metal figures in small but growing range. Price list available on request.
Hinchliffe Models Ltd, Meltham, Huddersfield HD7 3NX. Large and growing range of 'big' 25 mm figures suitable for wargames or dioramas. Price list available on request or illustrated Handbook £2.20.
Hinton Hunt Figures, 27 Camden Passage, Antique Village, Islington, London N1. Large range of 20 mm metal figures.
Miniature Figurines Ltd, 28/32 Northam Road, Southampton SO2 0PA. Currently make the largest range of 25 mm metal figures, and very easy to paint. Catalogue 25p.
Phoenix Model Developments Ltd, The Square, Earls Barton, Northampton. Small range of excellent 25 mm metal figures. Catalogue 16p.
Rose Miniatures, 15 Llanover Road, London SE18 3ST. Small range of 25 mm metal figures. Send SAE for price lists or 75p for full-colour catalogue.
Warrior Metal Miniatures, 23 Grove Road, Leighton Buzzard, Beds LU7 8SF. Small range of 25 mm figures. Catalogue 5p.

Note that not all the figures produced by the above firms are scaled precisely in relation to each other. Airfix figures are the smallest, followed by Hinton Hunt and Rose. Miniature Figurines, Garrrison, Warrior and Phoenix are all complementary, while Hinchliffe are slightly (and in some cases, considerably) larger. Airfix figures need washing in detergent and undercoating with polyurethane varnish before painting. Hinton Hunt and Warrior figures frequently require a great deal of cleaning up before painting, while Hinchliffe figures have lances and flag poles supplied separately. These have to be glued in place with epoxy resin. Prices vary from under a penny each for the Airfix figures to 10p or more per metal foot figure. The average price of

The Airfix OO/HO gauge set of British horse artillery figures and guns is one of the most useful sets in the range, since the shape of their helmets lends itself to conversion into a variety of types, including Bavarian or early Austrian infantry. You also get the guns and horses!

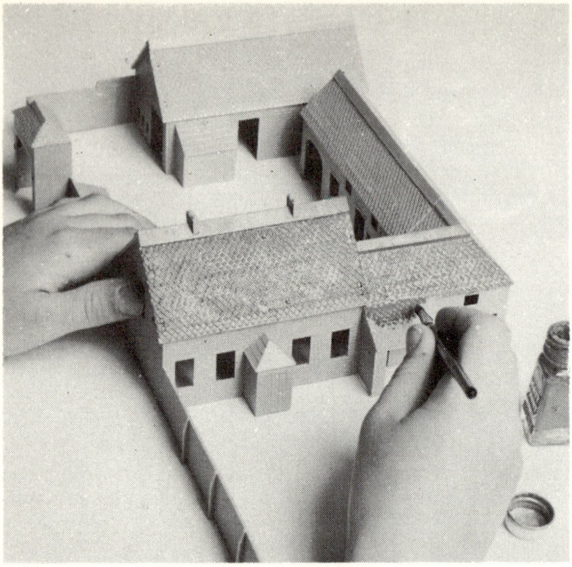

Apart from being an essential ingredient in any wargame of Waterloo, the Airfix La Haye Sainte farmhouse can also be cut up into three or more individual buildings to add variety to your terrain settings. If you leave the roof parts detachable figures can be placed inside with no problems.

a metal foot figure is 10 p. Prices for personality figures and standard bearers are greater in some ranges. In some ranges cavalry figures are moulded with the horses; in others they are separate. In many cases the cavalry figures are moulded together with the horse furniture, saddle etc. Reins, (from fuse wire) have to be added to Hinchliffe cavalry figures.

Bibliography

Two books are essential reading for wargamers specialising in this period. The first is David G. Chandler's epic *The Campaigns of Napoleon,* published by Weidenfeld and Nicholson, which is a detailed guide to every campaign and battle in which Napoleon played a part. It also contains valuable chapters on Napoleonic tactics and strategy, and is well illustrated with maps, diagrams and reproductions of contemporary paintings and prints.

The second is *Arms and Uniforms, The Napoleonic Wars* Vols I and II, by Liliane and Fred Funcken (Ward Lock Ltd). Between them these two volumes illustrate, in full colour, practically every uniform ever seen on a Napoleonic battlefield. The emphasis is on the French army and its allies but there are sections on the uniforms of Britain, Prussia, Austria, Russia, Spain, Portugal, Sweden and a host of minor states.

These three books are expensive but can be ordered through any public library if your finances won't stretch that far.

Apart from these, there are numerous books on specific Napoleonic battles and campaigns in the Knight's 'Battles for Wargamers' Series and Pan 'British Battles Series', both of which are relatively cheap. Similarly, there are various titles on Napoleonic uniforms in the Almark Publications range and the Osprey 'Men-at-Arms' series, all of which are available as paperbacks at an average price of £1.95. Shire Publications also include some useful titles in their inexpensive 'Discovering' series.

Wargames clubs

Model & Allied Publications Ltd, PO Box 35, Bridge Street, Hemel Hempstead, Herts, publish a directory of military modelling and wargaming clubs which is available from them for 30p (1977 price).